106 Advances in Polymer Science

Polymer
Characteristics

With contributions by
A. Grosberg, B. Hammouda, H.-W. Kammer,
E. T. Kang, J. Kressler, C. Kummerloewe
S. Nechaev, K. G. Neoh, K. L. Tan

With 62 Figures and 10 Tables

Springer-Verlag
Berlin Heidelberg GmbH

ISBN 978-3-662-14943-0   ISBN 978-3-540-47477-7 (eBook)
DOI 10.1007/978-3-540-47477-7

Springer-Verlag Berlin Heidelberg 1993
Softcover reprint of the hardcover 1st edition 1993

Library of Congress Catalog Card Number 61-642

Typesetting: Macmillan India Ltd., Bangalore-25

02/3020 - 5 4 3 2 1 0 Printed on acid-free paper

Editors

Table of Contents

Polymer Topology

Alexander Grosberg[1] and Sergei Nechaev[2]
[1] Institute of Chemical Physics, 117977 Moscow, Russia
[2] L.D. Landau Institute for Theoretical Physics, 117940 Moscow, Russia

We review the different mathematical approaches in the theory of topological constraints in statistical physics of long chain molecules.

1 Introduction: An Overview of Basic Definitions

Topology is the mathematical science dealing with statements of the following type: you can't peel an orange without breaking the skin or you cannot make an omelet without breaking eggs [1], you cannot comb the hair on a billiard ball, but you can do it on a torus, you cannot lasso a telegraph wire etc.

Properties of such as these play an important role for a lot of objects of different physical nature. Recently, some topological problems have been investigated in connection with quantum field and string theories, 2D-gravitation, polymer physics, etc. [2, 3, 4]. Polymers seem to be a natural, simple and very obvious example of systems with topological constraints.

From our point of view, the formulation of general topological problems in terms of polymer physics allows us:

– To increase our understanding of concrete physical properties of biological and synthetic polymer systems where the topological constraints play a significant role;
– To use a geometrically obvious image of the polymer with topological constraints as a model corresponding to the path integral formalism in quantum field theory.

In the present review, the main focus of attention will be on general problems concerning the topological properties of polymers presented as Brownian trajectories. We will not investigate the connection of the topology with the concrete chemical structure of polymers.

For a physicist, the polymer objects are attractive for many reasons. First of all, the joining together of monomer units to form chains essentially reduces all dynamic and equilibrium properties of the system considered. Moreover, due to this joining the behavior of polymers is determined by larger space-time scales than for low-molecular-weight substances and allows one to apply general theoretical methods such as perturbation theory, renormalization-group approach, conformal methods, etc. for investigating of polymer systems.

Great progress in the theoretical investigation of polymer systems is due to the combination of general methods of solid-state physics with approaches taking into account the chain-like structure of polymers.

It is well known that the chain-like structure of macromolecules causes the following peculiarities:

– the so-called "linear memory", i.e. the fixed position of each monomer unit along the chain;
– the low translational entropy connected with the fact that the independent motion of monomer units is prevented because of the presence of bonds;
– large space fluctuations because even a single macromolecule can be considered as a statistical system with many degrees of freedom (for more detail, see Ref. [5]).

At the same time, the above mentioned chain-like structure leads to the fact that different parts of polymer molecules fluctuating in space cannot go through each other without chain rupture. For the system of non-phantom closed chains, this fact means that only those space conformations that can be transformed continuously into one another are available (see Fig. 1). The adequate mathematical language for description of those physical effects is elaborated in the mathematical discipline called *topology*. That is why we also call the effects connected with chain uncrossability the *topological constraints*.

The topological state of the system of rings or uncrossable chains connected into a network coincides with that of the initial state and remains without any

Fig. 1. Left crossing can not be reduced to the right one by means of continuous deformation

changes in the course of brownian motion of polymer molecules. All the possible chain conformations form the classes of topologically equivalent states and the topological constraints in such systems are called strict.

The typical topological problem in polymer physics includes two inter-connected parts:

- To define the characteristic determining the topological state of macro-molecules. Such a characteristic is called *the topological invariant* [6]. It depends on the topological state of chains only and does not depend on the shape of the macromolecule.
- To determine the partition function, Z, i.e. the number of available chain conformations for fixed topological state.

Thus, it should be stressed that the *mathematical* topological theory inves-tigates, as a rule, the problems of classification of knots and links, the construc-tion of topological invariants, definitions of topological classes, etc.; whereas the fundamental *physical* problem in the theory of topological properties of polymer chains is the determination of the entropy, $S = \ln Z$ with the fixed topological state of chains. Both these problems are very difficult, but important.

It is noteworthy that strict topological constraints do not exist for systems of linear chains with free ends and do not affect the statistical properties of linear polymers; nevertheless, they significantly influence the dynamics of macro-molecules.

After these short explanation of basic definitions, it is quite easy to under-stand that because topological problems are very general they can be formulated easily using polymer language.

It is well known that the topological ideas are very easy to understand from the geometrical point of view but they are hard to formalize due to the non-local character of topological constraints.

Classifying the different approaches in this field, we will distinguish two basic groups of approaches called below "microscopical" and "phenomenological".

For microscopical approaches, we will ascribe the methods which allow one to investigate some simple, exactly solvable, basic models using quite rigorous mathematical ideas and to establish some general mathematical relations, in particular, concerning isomorphism between different physical problems.

Very few exactly solvable models of polymer systems with topological constraints have been described in the literature till now:

(1) Edwards–Frisch model of a polymer chain on a plane with a single removed point; this model will be discussed in Sect. 2.2.1;

(2) Some versions of the so-called model "chain in an array of obstacles", which geometrically can be considered in a certain sense as a generalization of Edwards–Frisch model; this will be considered in Sect. 2.2.3;

(3) The model of a smooth narrow strip. If the strip is narrow enough, i.e. if the inequality $d/a \ll 1$ is valid (where d is the strip width and a is the characteristic radius of curvature of the strip axis), then the Gauss linking number (Lk) characterizing the entanglement of strip borders with each other is the full topological invariant (due solely to the condition $d/a \ll 1$). Moreover, in the zeroth order approximation with respect to d/a-parameter, this topological invariant can be presented in the form:

$$Lk = Tw + Wr \tag{1}$$

i.e. using the differential-geometrical characteristics of a strip, namely its twisting, Tw, and writing, Wr. The Wr-value is determined by the spatial fold of the strip axis. An excellent feature of this result concerns its applicability in the closed circular DNA double helix theory. In this case, the borders of the strip can be identified with DNA strings, d/a means the ratio of double helix diameter to Kuhn segment – it is of order of 0.02 only, and this is why the relation $Lk = Tw + Wr$ is often considered to be exact for DNA. A lot of important results in DNA physics were obtained using this relationship. Among them are: double helix rigidity determination [7], super-helix model construction [8], closed circular DNA collapse analysis [9], etc.

Phenomenological approaches discussed in Section 4 do not contain the rigorous mathematical description of topological constraints. They are usually based on the geometrically clear conjectures on the character of polymer chain motion. This allows one to obtain the physically important results concerning the properties of complex polymer systems without the complete mathematical analysis. The theories of this type do not claim a high level of accuracy, but at present they remain very useful for the investigation of physically important properties of polymers with the topological constraints and play a very important phenomenological role.

2 Application of Non-Euclidean Geometry Ideas in Polymer Topology

2.1 Basic Models

The problem of investigation of polymer chain statistics without volume interactions entangled with an infinitely long string (in the 3D case) or with an obstacle (in the 2D case) was at first formulated by S.F. Edwards [10] and by S. Prager and H.L. Frisch [11] in 1967. These papers can be regarded as a corner stone in the building called "the theory of entanglements" in the statistical physics of polymers.

One can see [10, 13] that topological constraint even in this simplest case leads to the strong attraction of entangled chain to the obstacle or to repulsion for the unentangled chain.

Edwards' approach (Ref. [10]) is based on field-theoretic path-integral representation of the partition function $W_n(\vec{R}_0, \vec{R}_N, N)$ defining the probability density of the fact that end points of an N-link chain are placed at the points $\vec{R}_0$ and $\vec{R}_N$, respectively, and the chain turns n times around the string (the obstacle). The same problem in a slightly different way was considered by Prager and Frisch by using the combinatorial methods [11] and later by Saito and Chen by employing Fourier analysis [12].

From the physical point of view, it is clear that the topologically different states of the chain with fixed end points correspond to different numbers of turns around the string (or the obstacle). In other words, this number of turns, n, plays the role of topological invariant and it is proportional to the angle covering by the radius-vector connecting the obstacle and the point moving along the chain between its ends. Usually this value is called the "Gauss linking number", G. For closed chains, G takes values $2\pi n$ (the entanglement order, n, is an integer).

The partition function $W_n(\vec{R}_0, \vec{R}_N, N)$ has the following normalization condition:

$$\sum_{n=-\infty}^{\infty} W_n(\vec{R}_0, \vec{R}_N, N) = W(\vec{R}_0, \vec{R}_N, N) \tag{2}$$

It is well known that $W(\vec{R}_0, \vec{R}_N, N)$ satisfies the usual diffusion equation (boundary and initial conditions included):

$$\frac{\partial}{\partial N} W(\vec{R}, N) = \frac{a^2}{2d} \Delta W(\vec{R}, N)$$

$$W(\vec{R}, 0) = \delta(\vec{R}) \tag{3}$$

$$W(\infty, N) = 0$$

where d is the space dimension and $(a^2/2d)^{-1}$ is the diffusion coefficient (see, for instance, Ref. [13]).

2.2 Method of Conformal Transformations

2.2.1 A Polymer Chain Near the Single Obstacle

As was mentioned above, there are a lot of different ways of considering the Edwards–Frisch problem. However, from the methodological point of view and for the sake of a better clarification of non-euclidean geometry ideas for the description of topological constraints, we would like to present the method of conformal transformation.

The main idea is as follows. Let us consider the plane in which our chain is placed as a complex one, $z = x + iy$, $(z = z(x, y))$ and let us find the conformal transformation, $z = z(\zeta)$, of the plane z with the obstacle to the Riemann surface, $\zeta = \xi + i\eta$, which does not contain an obstacle (such a transformation means the transfer to the covering space). Due to the conformal invariance of Brownian motion[1], in the covering space ζ a random process will be obtained corresponding to the initial one on the plane z but without any topological constraints.

The reflection $z(\zeta)$ has the very simple form:

$$\zeta(z) = \ln(z) \tag{4}$$

Introducing the polar coordinates (ρ, θ) on the initial z-plane, we can rewrite Eq. (4) as follows:

$$\rho = \exp(\xi)$$
$$\theta = \eta + 2\pi n \tag{5}$$

where η changes in the interval $[0, 2\pi]$ and n means the number of copies of Riemann sheets. It is easy to see that, by means of the $z(\zeta)$-transformation, the origin of the z-plane (where the obstacle is placed) is transferred to the infinity of ζ-plane with a logarithmically branching point (see Fig. 2).

Taking into account that under the conformal transformation, the Laplace operator is transformed in the following way:

$$\Delta_z = \left|\frac{d\zeta}{dz}\right|^2 \Delta_\zeta, \tag{6}$$

where

$$\Delta_z = \frac{1}{\rho}\frac{\partial}{\partial\rho}\left(\rho\frac{\partial}{\partial\rho}\right) + \frac{1}{\rho^2}\frac{\partial^2}{\partial\theta^2}; \quad \left|\frac{d\zeta}{dz}\right|^2 = \frac{1}{|z'(\zeta)|^2}; \quad z'(\zeta) = \frac{dz}{d\zeta}, \tag{7}$$

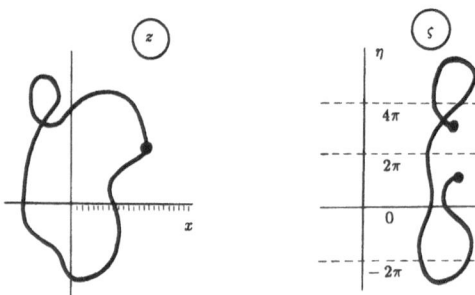

Fig. 2. Conformal transformation of z-plane with one point removed to the covering Riemann surface, ζ, without any peculiar points

---

[1] Conformal invariance of random walk means that after the conformal transformation this process will be random again.

we can write the diffusion equation for the distribution function in new coordinates remembering that the influence of topological constraint is reduced now to the boundary condition on the η-coordinate. Thus, we obtain:

$$\frac{\partial}{\partial N} W_0(\xi, \eta, N) = e^{-2\xi} \left(\frac{\partial^2}{\partial \xi^2} + \frac{\partial^2}{\partial \eta^2} \right) W_0(\xi, \eta, N) \tag{8}$$

with boundary conditions

$$0 \leq \eta < 2\pi; \quad W_0(\xi, n) = W_1(\xi, \eta - 2\pi) = \ldots = W_n(\xi, \eta - 2\pi n) \tag{9}$$

The solution of Eq. (8) with the conditions (9) has the following form (after transformation to (ρ, θ)-coordinates):

$$W_n(\vec{R}_0, \vec{R}_N) = \frac{1}{\pi N a^2} \exp\left(\frac{R_0^2 + R_N^2}{N a^2} \right)$$

$$\int_{-\infty}^{\infty} I_{|\lambda|}\left(\frac{2R_0 R_N}{N a^2} \right) e^{i\lambda(2\pi n - \theta_0)} d\lambda \tag{10}$$

where θ_0 is the angle distance between points $\vec{R}_0$ and $\vec{R}_N$ and $I_{|\lambda|}$ is the modified Bessel function.

It is noteworthy that a similar topological problem arises in 2 + 1-quantum field theory. Let us present the scattering amplitude α of quantum particle in terms of path integral with an arbitrary phase corresponding to each path. Then, for α we immediately obtain the expression:

$$\alpha \sim \sum_{\{C\}} \exp\left(i\hbar \int L dt \right) \tag{11}$$

If all phase trajectories by means of continuous deformation can be tranformed into one another, then the summation in Eq. (11) can be extended to all possible paths. However, if the phase space consists of topological non-isomorphic classes $\{C\}$, the summation in Eq. (11) extends only to paths from a given class and the problem of determination of the topological invariant arises.

The simplest examples of such systems in quantum physics are the interaction of the charged quantum particle with the Lagrangian $L = m\dot{\vec{r}}^2/2 + e/c\vec{A}\dot{\vec{r}}$ and the solenoidal magnetic field $\vec{A} = \vec{\varepsilon} \times \vec{r}/r^2$ (the Aharonov-Bohm effect) or the interaction of two anions in 2 + 1-field theory [14]. In both cases, the configurational space is the plane with one point removed.

2.2.2 The Non-Abelian Character of Entanglements

The natural generalization of the problem considered above is the calculation of the partition function for the chain entangled with many obstacles on the plane. At first sight, it seems that the approach presented above allows one to solve the problem easily. Actually, let us replace the function $\zeta(z)$ in Eq. (4) by the following:

$$\tilde{\zeta}(z) = \sum_{i=1}^{s} \ln(z - z_i) \tag{12}$$

where z_i are the coordinates of obstacles on the complex plane and s is the total number of obstacles. After conformal transformation to the covering space $\tilde{\zeta}$ by means of the function $\tilde{\zeta}(z)$, it can be seen that all obstacles are removed from any internal domain of the $\tilde{\zeta}$-plane and transferred to infinity. The topological invariant in that case will be the algebraic sum of the turns around each obstacle, which is a natural generalization of the Gauss linking number to the "many-obstacle entanglement".

However, the following problem arises here: for a system with two or more obstacles, it is possible to image closed trajectories entangled with a few obstacles together but not entangled with every one. In Fig. 3, the so called "Pochgammer contour" of such a type is shown. This type of conformations cannot be described using Gauss-type invariants.

Clarification of this point can be done using the concept of the "non-Abelian character of entanglements" for example of two obstacles. Let us denote each clockwise and counter-clockwise turn around the 1st obstacle by letters A and A^{-1} respectively. To the turns around the 2nd obstacle, we will attribute letters B and B^{-1} in a similar way. To each entanglement of any closed path with these two obstacles a "word" can be assigned consisting of a set of letters. Taking into account that $AA^{-1} = A^{-1}A = BB^{-1} = B^{-1}B = 1$, we can reduce each word to the minimal irreducible representation. For example, the word $W = AB^{-1}AAA^{-1}B^{-1}BA^{-1}B^{-1}$ can be transformed to the following irreducible form: $W = AB^{-1}B^{-1}$. It is easy to understand that the word $W \equiv 1$ corresponds only to unentangled contours. The entanglement in Fig. 3 corresponds to the irreducible word $W = A^{-1}BAB^{-1} \neq 1$. At the same time, the total algebraic number of turns (Gauss linking number) is equal to zero, so one would think that the path is not entangled with each obstacle. But this is incorrect! Thus, the non-Abelian character of topological constraints means that different entanglements do not commute: $[AB] = AB - BA \neq 0$.

Therefore, the principal difficulty connected with the application of Eq. (12) is due to the incompleteness of the Gauss invariant. So, the use of the Gauss invariant for adequate classification of topologically different states in many-chain systems is very problematic. Nevertheless, that approach was used repeatedly for consideration of such physically important question as the high-elasticity of polymer networks with topological constraints [15]. Unfortunately,

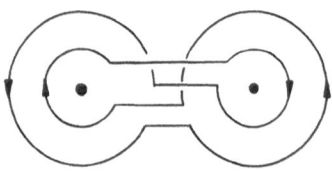

Fig. 3. Pochgammer contour entangled with two obstacles together but not entangled with either of them

in those works the correctness of the application of Gauss linking number was not considered.

2.2.3 Polymer Chain in the Lattice of Obstacles

The model "polymer chain in an array of obstacles" (PCAO) (see Fig. 4) combines the geometrical clarity of its image with the possibility to investigate the influence of entanglements on equilibrium and dynamic properties of polymers quite precisely.

It is noteworthy that for investigation of properties of real polymer systems with topological constraints it is not enough to be able to calculate the statistical characteristics of chains in the lattice of obstacles. It is also necessary to be able to compare any concrete physical system with the unique lattice of obstacles, which is a much more complicated problem than the first task. In this way, the model "polymer chain in an array of obstacles" is an intermediate between the microscopical and phenomenological approaches. The direct investigation of the PCAO-model was fulfilled in Refs. [16–25].

The word W consisting of a sequence of letters corresponding to different entanglements (introduced in Sect. 2.2.1) plays a role of full topological invariant for the PCAO-model. It is closely connected with the concept of the "primitive path" obtained by means of roughing of the microscopic chain trajectory up to the scale of the lattice cell and by exclusion of all loop fragments not entangled with the obstacles (Fig. 5).

In a number of papers [16–23, 25], the discrete variant of the PCAO-model is considered: the chain is modeled by a random walk on the lattice with spacing a and the topological constraints are placed on the dual lattice with period c.

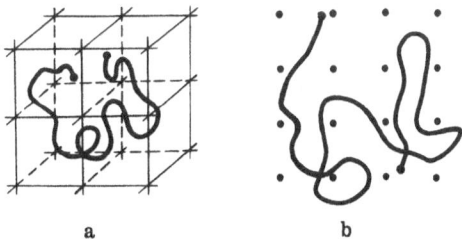

a   b

Fig. 4a, b. Chain in the lattice of obstacles: a) 3D-case; b) 2D-case

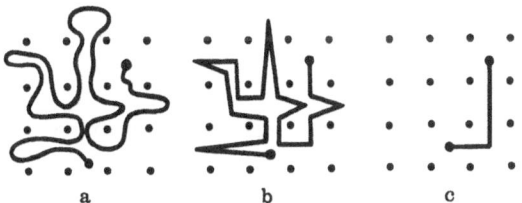

a   b   c

Fig. 5. Subsequent roughing of microscopic chain trajectory. The final state of the process coincides with the primitive path

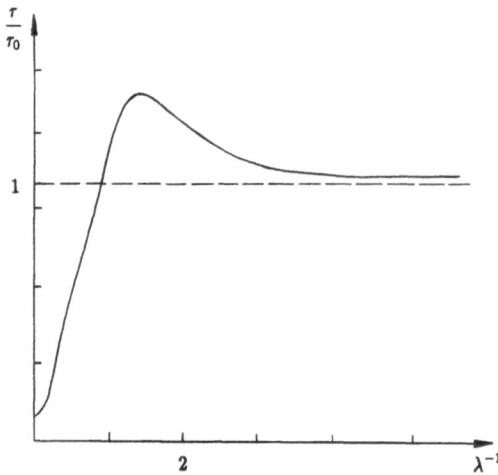

Fig. 6. Dependence of strain, $\bar{\tau}$, on relative deformation, λ, in Mooney-Rivlin coordinates ($\bar{\tau} = \tau/\tau_0$, λ^{-1})

The problem of determination of the partition function $Z(k, N)$ for the N-link chain having the k-step primitive path was at first solved in Ref. [17] for the case $a = c$ by application of rather complicated combinatorial methods. The generalization of the method proposed in Ref. [17] for the case $c > a$ was performed in Refs. [19, 23] by means of matrix methods which allow one to determine the value $Z(k, N)$ numerically for the isotropic lattice of obstacles. The basic ideas of the paper [17] were used in Ref. [19] for investigation of the influence of topological effects in the problem of rubber elasticity of polymer networks. The dependence of the strain τ on the relative deformation λ for the uniaxial tension $\lambda_x = \lambda_y = 1/\sqrt{\lambda}$, $\lambda_z = \lambda$ calculated in this paper is presented in Fig. 6 in Mooney–Rivlin coordinates (τ/τ_0, λ^{-1}), where $\tau_0 = vT/V_0(\lambda - 1/\lambda^2)$ represents the classical elasticity law [13]. (The direct Edwards' approach to this problem was used in Ref. [26].) Within the framework of the theory proposed, the swelling properties of polymer networks were investigated in Refs. [19, 23] and the $\tau(\lambda)$-dependence for the partially swollen gels was obtained [23]. In these papers, it was shown that the theory presented can be applied to a quantitative description of the experimental data.

We have presented only a few basic physical results obtained within the framework of the PCAO-model; the mathematical methods for investigating equilibrium and dynamic properties of a test polymer chain of different topology placed in the lattice of obstacles will be reviewed below.

2.2.4 Long Polymer Chain in the Lattice of Obstacles: Random Walk in Lobachevsky Space

The general definition of the problem is the following: inside the uniform regular lattice of obstacles with the elementary cell in a form of equal-sided triangle with the side of length c, is placed a polymer chain with N-links having a segment length a, and without volume interactions. The chain ends are fixed at points A and B (Fig. 5) The configuration of the primitive path connecting the chain ends defines the topological invariant and completely characterizes the topological state of the chain with respect to the lattice of obstacles (we assume that the chain does not produce entanglements with itself). In particular, the important topological invariant is the length of the primitive path, μ. The calculation of the partition function $W(\mu, N)$ is our main task.

Using the idea of conformal invariance again, let us find the conformal transformation $z = z(\zeta)$ of the plane z with a regular lattice of obstacles to the covering space $\zeta(\xi, \eta)$ which in every finite domain does not contain any obstacles. The construction of that transformation $z(\zeta)$ can be described in the following way: The elementary lattice cell of the z-plane (the ABC-triangle) is transferred in an equal-sided circle triangle lying on the Riemann surface ζ. It is obvious that by means of consecutive reflections of ABC-cell with respect to its sides and corresponding reflections in the covering space, all obstacles of z-plane will be transferred to the boundary of the open circle $|\zeta| < 1$ (see Fig. 7a, b). The group of motion in the covering space ζ in this case coincides with the group $SL(2, \mathbb{R})$. The open circle $|\zeta| < 1$ does not contain any of obstacles and is "expanded" with respect to the initial z-plane because the coordinates of chain ends on ζ-plane determine:

- The corresponding Euclidean coordinates on z-plane;
- The topological state of the chain trajectory on z-plane. In particular, the closed contours on the ζ-plane correspond only to *unentangled* closed contours on the z-plane.

It can be seen that our problem is now reduced to the solution of the diffusion equation

$$\frac{\partial}{\partial N} W(\zeta, N) = \frac{a^2}{4} \frac{1}{|z'(\zeta)|^2} \Delta W(\zeta, N) \tag{13}$$

in the ζ-circle without any topological constraints!

The explicit form of the transformation $z(\zeta)$ was found in Ref. [24]. We will not reproduce the calculations obtained from there, but only comment on the result. The structure obtained by the transformation $z(\zeta)$, called the "modular

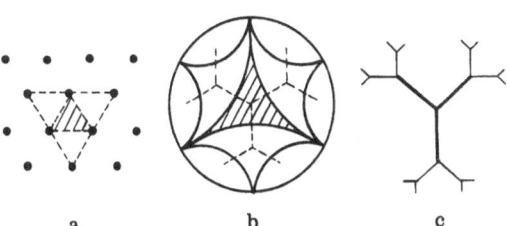

Fig. 7a–c. Conformal transformation of the plane with obstacles (a) to the modular figure with Lobachevsky-metric (b) and its topological structure (c)

group", is defined via relation:

$$\zeta \to \frac{a\zeta + b}{c\zeta + d}; \quad ad - bc = 1 \tag{14}$$

and its topological structure coincides with the Cayley tree (see Fig. 7c). Taking into account that the group of modular transformation is contained in the group of motions of Lobachevsky plane as its discrete subgroup and neglecting the metrics changing on scales of order of curvilinear triangle, one can write for $|dz/d\zeta|^2$ the expression which represents the metrics of Lobachevsky plane:

$$\left| \frac{dz(\zeta)}{d\zeta} \right|^2 \simeq \frac{3\beta^2 c^2}{|1 - \zeta\zeta^*|^2} \tag{15}$$

where

$$\zeta = re^{i\psi}; \quad \zeta^* = re^{-i\psi}; \quad \beta = \pi^{-1/3} B^{-1}(1/3, 1/3)$$

After substitution of Eq. (15) into Eq. (13) we obtain the random walk on the Lobachevsky plane in the Poincaret model [27]. Introducing the non-Euclidean distance $\mu = \frac{1}{2}\ln(1 + r)/(1 - r)$, it is easy to obtain the following asymptotic solution for $\mu/c \gg 1$

$$W(\mu, N) \simeq \frac{1}{4\pi N a^2} \exp\left(-\frac{1}{4\lambda^2 N}(\lambda^2 N + 2\mu/c)^2 \right) \tag{16}$$

where $N = L/a$ and $\lambda^2 = a^2/(3\beta^2 c^2)$ – is the scalar curvature of the Lobachevsky plane.

Equation (16) defines the probability density of the fact that two *concrete* end points of N-link chain are connected by the primitive path of length μ. The probability density for an *arbitrary* end points having the distance equal to μ is defined by:

$$\tilde{W}(\mu, N) = W(\mu, N) \sinh(2\mu/c) \tag{17}$$

Unfortunately, for the investigation of random walk statistics in the regular 3D lattice of obstacles the approach based on the idea of conformal transformations cannot be applied. Nevertheless, due to the analogy established in the 2D-case it is naturally to suppose that between random paths statistics in the 3D lattice of uncrossable strings and the free random walk in Lobachevsky space the similar analogy remains. Let us present below some arguments confirming that idea.

The coordinates of chain ends in the initial $z^{(3)}$-space do not determine the topological state of the chain completely. There is an exponentially large number of chain configurations with different primitive paths (i.e. with different topological states). Let us introduce the *generalized coordinates* $\zeta^{(3)}$ determining the Euclidean coordinates of the chain in the initial z-space as well as the topological invariant of the chain. The random walk in the space of generalized coordinates $\zeta^{(3)}$ with the defined boundary conditions corresponds to the initial problem. The $\zeta^{(3)}$-space (hyper-space) cannot be enclosed in the usual euclidean space. One can easily understand that for the discrete case the Cayley tree, being an exponentially growing structure, is enclosed "densely" (without gaps and selfintersections) in the hyperboloid. That hyperboloid plays a role of the space of generalized coordinates for the 3D continuous PCAO-model. According to the terminology adopted in cosmology, let us call it "3-pseudo-sphere" or Lobachevsky space [28].

Analogous to Eq. (13) we can write the equation for the 3D-case:

$$\frac{\partial}{\partial N} P^{(3)}(\mu, N) = \frac{a^2}{9} \beta^2 \tilde{\Delta} P^{(3)}(\mu, N) \tag{18}$$

where $\tilde{\Delta}$ is Beltrami–Laplace operator

$$\tilde{\Delta} = \frac{1}{\sqrt{\det \| g^{(3)} \|}} \frac{\partial}{\partial x^i} \left(\sqrt{\det \| g^{(3)} \|} \, g^{ik} \frac{\partial}{\partial x_k} \right) \tag{19}$$

and the metric tensor equals

$$\| g^{(3)} \| = \left\|\begin{array}{ccc} 1 & 0 & 0 \\ 0 & c^2/4 \sinh(2\mu/c) & 0 \\ 0 & 0 & c^2/4 \sin(2\mu/c) \sin^2 \theta \end{array}\right\| \tag{20}$$

Lobachevsky-space metric is

$$ds^2 = d\mu^2 + \frac{c^2}{4} \sinh^2(2\mu/c)(d\theta^2 + \sin^2 \theta \, d\phi^2) \tag{21}$$

The solution of Eq. (18) yields [27]

$$W^{(3)}(\mu, N) = \left(\frac{3}{2\pi N a^2} \right)^{3/2} \exp\left(-\frac{4a^2 N}{9\beta^2 c^2} - \frac{9\mu^2}{4\beta^2 a^2 N} \right) \frac{2\mu/c}{\sinh(2\mu/c)} \tag{22}$$

In Ref. [23], the value $W^{(3)}(\mu, N)$ was calculated exactly for the lattice model. However, the method proposed there did not allow one to obtain an analytic expression for the $W^{(3)}(\mu, N)$-function. Comparing the values obtained from Eq. (22) with the numerical data of Ref. [23], one can make a conclusion that Eq. (22) can be regarded as a good interpolation for the exact solution in Ref. [23].

The approach presented here allows us to calculate such a physically important observable quantity as the gyration radius of a closed chain unentangled with obstacles. Omitting the nonuniversal numerical coefficients depending on the space dimensionality and coordinational number of the lattice of obstacles, the following scaling relation obtained in Refs. [18, 22] is fulfilled:

$$\langle R_g^2 \rangle \sim acN^{1/2} \tag{23}$$

Let us remind that for a usual non-restricted random walk we would obtain the well-known equation (see, for instance, Ref. [13]):

$$\langle R_g^{0^2} \rangle \sim a^2 N \tag{24}$$

Comparing Eqs. (23) and (24), it can be seen that the unentangled closed path in the lattice of obstacles is strongly contracted with respect to the usual Gaussian one.

3 Algebraic Invariants of Knots and Links, and Non-Abelian Field Models

3.1 Alexander Polynomials as a Tool for Numerical Investigations of Polymers with Topological Constraints

We have seen in the previous sections, that the theory of a non-phantom polymer chain is not very simple from the technical point of view even in the case when topological constraints can be described by the Gauss linking number. It has been explained that the principal difficulty is connected with the fact that the Gauss linking number is a weak topological invariant due to non-Abelian properties of entanglement. Its application is very restricted in strong entangled systems because of its incompleteness.

In particular, the most obvious topological question, concerning the probability of knotting during random closure of polymer chain, cannot be answered using the Gauss invariant.

Great progress in this field is connected with the works of Vologodskii et al. [29–33] and others [34–38]. In these works, the algebraic polynomials were used for the topological state identification of closed polymer chains generated by Monte-Carlo methods. These algebraic invariants, such as Alexander polynomials [6], are much stronger than the Gauss invariant. This approach based on Alexander polynomials proved to be extremely fruitful for computer simulations, and the main part of our modern knowledge on knots and links was obtained with the help of this method. Among the most important results there are the following:

– The dependence $p(N)$ of the chain self-knotting probability p is determined as a function of chain length N by the random chain closure [29–31, 34]. In a recent work [36], the simulation procedure was extended up to the chains of order $N \simeq 2000$, where for trivial knot formation probability the exponential asymptotics of the type

$$p_0 \sim \exp(-N/N_0)$$

is obtained for chains in good and θ-solvents;

– The knot formation probability p was investigated as a function of swelling ratio $\alpha(\alpha < 1)$ where $\alpha = S/S_0$, $\langle S^2 \rangle$ is the mean-square gyration radius of the closed chain and $\langle S_0^2 \rangle = Na^2/12$ is the same for unperturbed ($\alpha = 1$) chain (see Fig. 8) [29–31, 33, 35, 37]. It has been shown that this probability

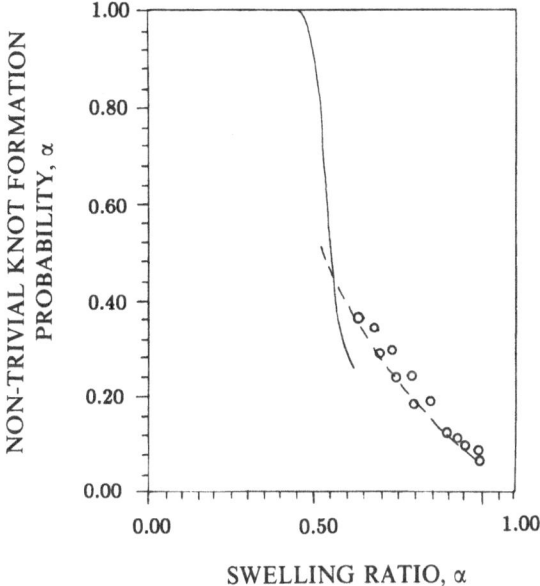

Fig. 8. Dependence of non-trivial knot formation probability, p, on swelling parameter, α, in globular state. *Points*-data from Ref. [29]; *dashed line* – approximation in weak compression regime; *solid line* – approximation based on concept of crumpled globule (Eq. (53))

decreases sharply when a coil contracts from the swollen state with $\alpha > 1$ to the Gaussian one with $\alpha = 1$ [33, 37] and especially when it collapses to the globular state [29, 35].

- It has been established that in region $\alpha > 1$ the topological constraints are screened by volume interactions almost completely [33].
- It has been shown that two chains unentangled with each other, even without volume interactions in the coil state, repulse each other as impenetrable spheres with radii of the order S_0 [29, 34].

The so-called Jones polynomials [38] are even more strongly invariant than the Alexander ones. However, their calculation requires far more computer capacity: calculation of an Alexander polynomial takes in the order $O(l^3)$ operations, where l is the number of selfintersections of contour projection on the plane; on the other hand, the calculation of a Jones polynomial takes in the order $O(e^l)$ operations. This is why the existing attempts to use Jones polynomials in computer experiments with ring polymers have not been successful as yet. Nevertheless, the construction of algebraic polynomial invariants of knots and links seems to be of great importance in principle, and we shall consider it in the next section.

3.2 Reidemeister Moves, State Model for Construction of Algebraic Invariants and Yang–Baxter Relations

We would like to describe in this section the very beautiful idea proposed by L.H. Kauffman for the analytical construction of powerful polynomial invariants of knots and links.

Below, we shall consider only the case of knots because the link invariants could be constructed in the similar way.

Let K be a knot embedded in 3D-space. First of all, we project it onto a plane and obtain the 2D-image in the so-called general position. It means that only pair crossings can be in this projection. Then, for each crossing we define the passages, i.e. parts going below and above in accordance with its natural positions in 3D-space.

For the knot plane projection with defined passages, the following Reidemeister theorem is valid [39]: different knots (or links) are topologically isomorphic to each other if they can be transformed continuously into one another by means of a sequence of simple local Reidemeister moves of types 1, 2 and 3 (see Fig. 9). Two knots are called *regular isotopic* if they are isomorphic with respect to the last two types of moves (2 and 3); if they are isomorphic with respect to all types of Reidemeister moves, they are called *ambient isotopic*. As can be seen from Fig. 9, a Reidemeister move of type 1 leads to the cusp creation on chain projection. At the same time, it is noteworthy that all real 3D-knots (links) are of ambient isotopy.

Now, when we have formulated the Reidemeister theorem, we can describe the construction of a powerful polynomial "bracket" invariant proposed by L.H. Kauffman [40, 41]. This invariant can be introduced as a certain partition function, which is the sum over the set of some formal degrees of freedom.

Let us consider 2D-projection of our knot as a certain irregular lattice. Crossings of projection are lattice vertices. We turn all these crossings into a standard position where two parts of chain projection near the crossing are normal to each other and form the angles of $\pi/4$ with the x-axis. It can be proved that the result does not depend on the method of this standardization. Now, we

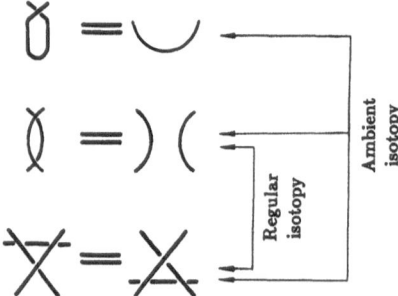

Fig. 9. Illustration of Reidemeister moves

have two kinds of vertices on our lattice

$\times$ *a)*

and

$\times$ *b)*

The next step of invariant construction is the introduction of two-state "spins" in all vertices. The states of each "spin" correspond to two possible ways of *vertex splitting*

$)($ *vertical*

and

$\times$ *horizontal*

To each state of spin, i.e. to each way of splitting, we attribute the following statistical weights:

– A to vertical splitting and B to horizontal splitting if the vertex is of type *a)* and,
– B to vertical splitting and A to horizontal splitting for the vertex of type *b)*.

For the system with N vertices, i.e. passages of chain projection, there exist 2^N different microstates, each of them representing a set of splittings of all N vertices. Each microstate, S, corresponds to lattice disintegration to the system of disjoint and nonselfintersecting circles. The number of these circles for microstate S we will call $|S|$.

Then, the Kauffman polynomial can be written as follows:

$$\langle K \rangle = \sum_{\{S\}} d^{|S|-1} A^i B^j, \tag{25}$$

where $\sum_{\{S\}}$ means summation over all possible microstates of the knot (or link), i and $j = N - i$ are the numbers of vertices with weights A and B, respectively, in the given realization of microstate S.

The partition function of the system described above represents the polynomial in A, B and d values. Just this function, for some special choice of relations among weights A, B and d, is the topological invariant of regular isotopic knots. Let us prove this fact and derive the necessary relation among A, B and d values.

We shall use the following diagrammatic rules. Denote with $< \ldots >$ the statistical weight of the knot or its part; the $\langle K \rangle$ value equals the product of all the weights of the knot parts. Using Eq. (25), it is easy to check the following identities:

$$\langle K \cup O \rangle = d \tag{26}$$

$$\langle \times \rangle = \langle)(\rangle A + \langle \times \rangle B \tag{27}$$

$$\langle \times \rangle = \langle)(\rangle B + \langle \times \rangle A \tag{28}$$

where Eq. (26) defines the invariant for composition of knot K and the trivial ring; Eqs. (27) and (28) correspond to weights of splittings defined above. These diagrammatic rules are well defined only for a fixed "boundary condition". Let us suppose that, by convention, the polynomial for trivial ring is equal to unity; i.e.:

$$\langle 0 \rangle = 1 \tag{29}$$

The proof is based on direct checking the invariance of the $\langle K \rangle$-value with respect to Reidemeister moves of types 2 and 3. For the Reidemeister move of type 2 we have:

$$\langle \,\rangle = \langle \,\rangle ABd + \langle \,\rangle A^2 + \langle \,\rangle B^2 +$$
$$+ \langle \,\rangle AB = \langle \,\rangle (ABd + A^2 + B^2) + \langle \,\rangle AB \tag{30}$$

Therefore, the invariance with respect to the Reidemeister move of type 2 can be obtained immediately if the values of A, B, and d obey the following system of equations:

$$AB = 1$$
$$\tag{31}$$
$$ABd + A^2 + B^2 = 0$$

It can be also checked that the condition of invariance with respect to the Reidemeister move of type 3 does not violate Eqs. (31). Actually,

$$\langle \,\rangle = \langle \,\rangle B + \langle \,\rangle A$$
$$= \langle \,\rangle B + \langle \,\rangle A = \langle \,\rangle \tag{32}$$

Thus, the theorem is proved.

Equation (31) can be converted into the form

$$B = A^{-1}, \quad d = -A^2 - A^{-2} \tag{33}$$

and it means, that the Kauffman invariant (25) can be presented as a Laurent polynomial in A value.

The invariant of ambient isotopy of oriented knot or link is defined by:

$$f[K] = (-A)^{-3w(K)} \langle K \rangle \tag{34}$$

where $w(K)$ is the twisting of the knot (link), or, in other words, the sum of signs of all crossings defined by the convention:

$$\times +1 \qquad \times -1$$

The state model and bracket polynomials introduced by L.H. Kauffman seem to be very special because they explore only the peculiar geometrical rules such as summation over all possible knot (link) splittings with simple defined weights. But L.H. Kauffman also showed that bracket polynomials are strongly connected with the Jones polynomials [38]. The substitution $A = t^{-1/4}$ converts

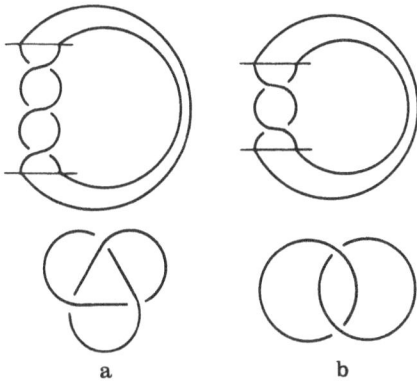

Fig. 10a, b. Representation of knot (a) and link (b) as braids

the bracket polynomial in A-value to the original Jones polynomial in t-value [41].

The Jones polynomial invariants for knots and links were discovered first by V. Jones during his investigation of topological properties of braids. Actually, each knot (link) can be presented in form of a braid (see Fig. 10), but this correspondence is not one-to-one. There exists a topological theorem proved by Markov (see, for example, Ref. [40]) which establishes the necessary conditions for the equivalence of knot (link) invariant and corresponding invariant of braids. An extremely powerful approach by Jones made possible the establishment of the equivalence of relations among generators of a braid group and the so-called Yang-Baxter relation which geometrically resembles the Reidemeister move of type 3. The Yang-Baxter relation plays an exceptionally important role in statistical physics of integrable systems (such as ice, Potts, $O(n)$, 8-vertex, quantum Heisenberg models [43]) and means the condition which is necessary for commutation of transfer matrices [44–46].

The attempt to apply the Kauffman invariants of regular isotopy to investigating statistical properties of random walks with topological constraints in a thin slit was made in a recent work [47].

To emphasize the broad region of applicability of the system described in this section, we would like to stress the following fact. Recently, in Refs. [48, 49] during investigation of 3D-quantum field theory with Chern-Simon's action a strong connection was established between expectation values of Wilson lines with non-trivial topology and partition function determining the polynomial invariant of the knot or link.

4 Phenomenological Approaches for Description of Physical Properties of Entangled Polymers

The generalization of the microscopic approaches for description of real many-chain polymer systems such as networks, concentrated solutions and melts requires a set of uncontrolled conjectures which lowers the value of the results obtained.

In that connection, it is of special importance to investigate approximate and qualitative approaches based on intuitive geometrical ideas concerning chains as uncrossable "ropes". Although these approaches do not allow complete mathematical analysis, they are useful from the empirical point of view.

4.1 The "Polymer Chain in a Tube" Model and Similar Ones

The first phenomenological model for description of polymer dynamics in concentrated solutions and melts was proposed in 1971 by P.de Gennes [50]. In this classical work, it was assumed that due to entanglements, the chain motions in the direction normal to the chain contour are blocked up and only tangential ones are possible. This kind of chain motion in the effective tube was called *reptation*. In the absence of external fields, the chain can escape from the tube by either of the free tube ends.

In the course of investigation of polymer dynamics, all topological states of chains are equivalent due to the presence of their free ends and it is not necessary to construct the topological invariant. However, for every instantaneous chain configuration, its topological state is definite and is determined by the topological invariant – the primitive path.

The development of the "polymer chain in a tube" model by M. Doi and S.F. Edwards [51] allowed them to describe on the molecular level a set of hydrodynamic and viscoelastic properties of concentrated polymer solutions and melts.

The "polymer chain in a tube" model and its modifications are widely used for investigation of the rubber elasticity of polymer networks with topological constraints. The detailed review on that problem one can find in the monograph [51].

In general, it should be said that the "polymer chain in a tube" model describes a number of physical properties of many-chain polymer systems in correspondence with experimentally observed data [52]. However, a complete analysis of the rubber elasticity and swelling properties within the framework of that model shows that none of the theories can claim an exact agreement with *all* experimental data.

The obvious difficulties appearing in the description of topological constraints within the frame work of the "chain in a tube" model are the following:

– It is not clear what the tube diameter is and what is its dependence on strain;
– The tube models can be applied to the investigation of the networks only if the number of chain segments, N, is far more than N_e – the characteristic distance between neighboring entanglements along the chain, expressed in the number of segments. As a rule, the values of N_e lie in the range of $50 - 300$ [13] and the real values of N do not exceed 1000. This means that the inequality $N \gg N_e$ is not always fulfilled;

– For the description of statistical properties of melts of ring polymers the notion of the "tube" does not make sense.

Thus in some respects the "polymer chain in a tube" model is imperfect and the more detailed "polymer chain in the lattice of obstacles" model mentioned above is preferable.

It should be noted that for investigation of the statistical or relaxational properties of solutions of polymers with fixed topology (polymer rings, for instance) the dynamic Monte-Carlo method [53] and its modifications [54] are widely used. These methods do not require the use of any invariants and are closest to the true chain motion.

4.2 The Quasi-Knot Concept for Polymers with Topological Constraints

The strict mathematical definition of a knot can be formulated for a closed (or infinite) contour only. Nevertheless, everyday experience tells us that a rope can be knotted and, moreover, often does so spontaneously and, as a rule, at unsuitable moments. Due to the last fact, it seems very attractive to construct somehow a non-rigorous notion of a *quasi-knot* for description of linear chains with free ends. This approach was discussed first in 1973 by I.M. Lifshits and one of the present authors [55] for the globular state of the chain. The main conjecture is the following: in the globular state, the distance between chain ends $\sim N^{1/3}$ is much smaller than the chain contour length ($\sim N$); therefore, the topological state of closed contour, consisting of the chain backbone and the straight end-to-end part, depends slightly on the positions of the chain ends. Just the knot of this closed contour can be considered as a quasi-knot of the linear chain.

It would be interesting to find out, using computer simulations, if this quasi-topological state of the linear globular chain is practically independent of the positions of the chain ends.

The quasi-knot concept seems to be most obvious in the case when the knot can be considered as a local one, i.e. when it has its own internal scale much smaller than the chain scale as a whole. However, for chains in the coil state the computer experiments do not indicate the existence of such an internal knot scale (for the globular state the detailed experiments are absent).

Nevertheless, A.N. Semenov in his works [56–59] have used the concept of local quasi-knot for Gaussian coil state, because the end-to-end distance in this state ($\sim N^{1/2}$) is still less than the chain length ($\sim N$). Within the framework of this approach, the authors of Refs. [56, 57] have shown that the topological constraints do not influence the dynamics of the whole polymer coil suspended in dilute θ-solutions but lead to the essential slowing-down of the internal coil excitation modes. In accordance with Refs. [56–59], the maximum relaxation time of a coil is limited by the diffusion of the local knots along the chain, i.e. the knots localized on a finite (when N tends to infinity) scale of order of one segment, and this time if of the order

$$\tau \sim N^3 \tag{35}$$

contrary to the classical Zimm result for a phantom chain $\tau \sim N^{3/2}$.

The most interesting result of Refs. [58, 59] concerns the crossover regime between dilute and semidilute regions of polymer θ-solution. The author shows that in this crossover regime there exists the critical concentration c^* corresponding to the appearance of an infinite cluster of entangled with each other macromolecules. It is also shown that near this critical concentration the relative viscosity η_r of the θ-solution has a scaling form:

$$\eta_r \sim N^{1.5}(c/c^* - 1)^6 \tag{36}$$

where $0 < c/c^* - 1 \ll 1$ and the self-diffusion coefficient D of the coil decreases exponentially according to the law

$$D \sim \exp(- c/c^*). \tag{37}$$

4.3 Replica-Trick Application for Non-Phantom Networks

In this section, we will briefly describe the new approaches to the topological problems in polymers proposed independently by S.V. Panykov [60] and by P. Goldbard and N. Goldenfeld [61].

To our mind, the main idea in these papers can be explained as follows; It is clear that the fixed topological structure of a polymer network is a typical example of the so-called quenched disorder. This structure is formed during the network preparation and cannot be changed later without network destruction. Because of the strict topological constraints on chain conformations, the full phase space of the network is divided into separated domains – just like the multi-valley structure of the spin glass phase space. Every domain corresponds to the sub-space of chain conformations with the fixed value of the topological invariant. The methods of theoretical description of the systems with quenched disorder is highly developed at present, especially in connection with spin glass theory [62, 63].

The key point of these methods is the concept of self-averaging: in accordance with this concept, the observable value of an additive function, for example – of the free energy, F, of macroscopic sample of spin glass, practically coincides with the averaged value over the ensemble of disorder realizations:

$$F_{obs} = \langle F \rangle_{av}. \tag{38}$$

The phenomenon of self-averaging takes place for the systems with sufficiently small long-range correlations: only in this case one can consider F as a sum of contributions from different volume elements, containing different and statistically independent realizations of disorder (for more details see Ref. [63]).

Therefore, the central technical problem is the calculation of averaged value of $F = -\ln Z$; this can be done using the so-called replica trick:

$$F_{obs} = \langle F \rangle_{av} = -\langle \ln Z \rangle_{av} = -\lim_{n \to 0} \frac{\langle Z_{av}^n \rangle - 1}{n} \tag{39}$$

This replica-trick method was used in the Refs. [60, 61] for the polymer network free energy calculation. For averaging of Z^n-value over the ensemble of realizations the probability distribution should be chosen. The authors of Refs. [60, 61] have used the condition of the thermodynamical equilibrium for the Gibbs probability distribution corresponding to the conditions of network preparation. To our mind, it is a beautiful idea but it should be considered more deeply because not all the types of networks can be described in such a way – many networks cannot be prepared under any equilibrium conditions. Using some additional "tube-like" approximations, the authors have obtained rather simple results for network elastic constants and for some other parameters.

However, in our opinion, the approach described in this section has some shortcomings. They are not connected with the replica trick application and with the rough approximations (which could be improved). The main objections concern the fundamental hypothesis of free energy self-averaging of a non-phantom network. This hypothesis can be criticized because of the existence of some long-range topological correlations in the network. As a whole, the problem remains open and one can suppose that some conditions of self-averaging principle validity would be formulated in an explicit form.

4.4 The Crumpled Globule State Concept for Estimation of Knot Formation Probability

Let us return to Fig. 8, where the knot formation probability p is plotted as a function of the swelling ratio, α, in the globular region ($\alpha < 1$). It can be seen that in the compression region, especially for $\alpha < 0.6$, data of numerical simulation are absent. It is difficult to obtain such data because of the restricted capacity of computers. Really, it is necessary to calculate the Alexander polynomial for each generated closed contour. As mentioned above, it takes in the order of $O(l^3)$ operations. This value is too large for the dense chain state because the denser the system is, the more selfcrossings l should be in the projection.

Let us present now the theoretical estimations of the non-trivial knot formation probability $p(\alpha)$ in dense globular state ($\alpha < 0.6$) based on the concept of "crumpled globule" proposed in Refs. [64, 65].

It is easy to understand that the trivial knot formation probability under random linear chain closure, $q(\alpha) = 1 - p(\alpha)$, can be defined by the relation:

$$q(\alpha) = \frac{Z(\alpha)}{Z_0(\alpha)}, \tag{40}$$

where $Z(\alpha)$ is the partition function of an unknotted closed chain with volume interactions for fixed value of the swelling parameter, α, and $Z_0(\alpha)$ is the same of "shadow" chain without topological constraints but with the same volume interactions. Both these partition functions can be estimated using a mean field approximation. To do so, let us write down the classical Flory-type representation for the free energy of the chain with a given α value (in equations below we suppose T to be equal to unity):

$$
\begin{aligned}
F(\alpha) &= -\ln Z(\alpha) &= F_{int}(\alpha) + F_{el}(\alpha) \\
F_0(\alpha) &= -\ln Z_0(\alpha) &= F_{int}(\alpha) + F_{el,0}(\alpha) \\
F_{el}(\alpha) &= -S(\alpha), \; F_{el,0}(\alpha) &= -S_0(\alpha)
\end{aligned}
\tag{41}
$$

Here, the contributions $F_{int}(\alpha)$ from volume interactions to the free energies of the unknotted and shadow chain of the same density (or α) are the same. Therefore, the only difference concerns the elastic part of free energy, or, in other words, the conformational entropy. Thus, Eq. (40) can be represented in the form:

$$
q(\alpha) = \exp(-F(\alpha) - F_0(\alpha)) = \exp(S(\alpha) - S_0(\alpha))
\tag{42}
$$

According to Fixmann calculations [66], the entropy of a phantom chain $S_0(\alpha) = \ln Z_0(\alpha)$ in the region $\alpha < 1$ can be written in the following form:

$$
S_0(\alpha) \simeq -\alpha^{-2}
\tag{43}
$$

Now, we can write down the functional expression for the non-trivial knot formation probability $p(\alpha)$ depending only on the thermodynamic characteristic of the polymer chain, $S(\alpha)$. Combining Eqs. (42) and (43), we obtain the following relation:

$$
p(\alpha) = 1 - \exp(\alpha^{-2} + S(\alpha))
\tag{44}
$$

Therefore, the nontrivial part of our problem is reduced to estimation of the entropy of a strongly contracted closed unknotted ring. However, it was shown in Refs. [64, 65] that such a ring under strong contraction forms the so-called crumpled globule state with a unique fractal structure of line representing the chain fold. The entropy loss caused by crumpled globule state formation was estimated in Refs. [65, 67] using qualitative arguments and the result is as follows:

$$
S(\alpha) \simeq -\frac{1}{N_e}\alpha^{-6}
\tag{45}
$$

In the region of our interest ($\alpha < 0.6$), the α^{-2}-term in Eq. (44) can be neglected in comparison with α^{-6}. Therefore, we obtain finally the probability estimation in the form:

$$
p(\alpha) = 1 - \exp\left(-\frac{1}{N_e}\alpha^{-6}\right)
\tag{46}
$$

Let us explain now the concept of the crumpled globule state and estimation (45).

After a temperature decrease, the formation of the globular structure is thermodynamically favorable. Supposing that the final state can be described in virial expansion we introduce as usual two- and three-body interaction constants: $B = b \frac{T-\theta}{\theta} < 0$; and $C = const > 0$. However, in addition to the volume interactions we should also take into account non-local topological constraints having a repulsive character. In this connection, we express our main conjecture: the topological constraints lead to special non-trivial fractal properties of line representing the chain trajectory in the globule. Let us describe the structure.

It is well-known that in a poor solvent there exists some critical chain length, g^*, depending on temperature and energy of volume interactions, so that chains longer than g^* collapse. With regard to sufficiently long chains, we shall define these g^*-link parts as crumples of minimal scale.

Let us consider the part of a chain with several block monomers, i.e. crumples. This part of the chain should collapse in itself, i.e. form the crumple of the next scale, if other ones do not interfere with it. The chain of such collapsed subglobules, or crumples, should also collapse in itself, up to the chain as a whole (see Fig. 11). This procedure is completed when all initial links are united into one crumple on the largest scale. It is noteworthy that the line representing the chain trajectory obtained by the procedure described above resembles the well known Peano curve.

It seems at the first sight that due to space fluctuations all crumples could penetrate each others with loops, destroying the self-similar scale-invariant crumpled structure described above.

However, we can show that, if the chain length in a crumple of an arbitrary scale exceed N_e, the crumples coming in contact do not mix with each other and remain segregated in space.

Let us clarify the last statement in detail.

Because of the fact that the topological state of the chain part in every crumple is fixed and coincides with the state of the whole chain, which is unknotted, this chain part can be presented as an unknotted ring. Others chain

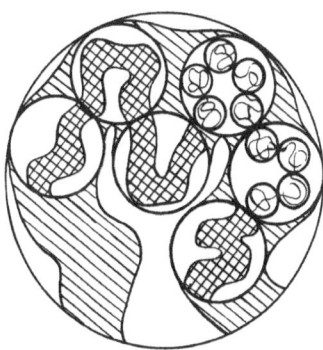

Fig. 11. Self-similar, segregated on all scales, structure of a crumpled globule

parts (other crumples) play a role of an effective lattice of obstacles surrounding the test ring. However, it was mentioned in Sect. 2 (see Eq. (23)) that the M-link ring chain without volume interactions not entangled with any of obstacles has the size

$$R^{(0)}(M) \sim aM^{1/4} \tag{47}$$

As $R^{(0)}$ is the size of the equilibrium chain part in the lattice of obstacles, the entropy loss for a ring chain, S, as a function of its size, R, reaches its maximum for $R \simeq R^{(0)}$ and further chain swelling for values of R exceeding $R^{(0)}(M)$ is entropically unfavorable. At the same time, in the presence of excluded volume, the following obvious inequality must be fulfilled:

$$R(M) \sim aM^{1/3}. \tag{48}$$

In connection with the obvious relation

$$R(M) > R^{(0)}(M) \tag{49}$$

the swelling of chains in crumples due to their mutual interpenetration with the loops does not lead to entropy gain and, therefore, does not occur in the system with limited density. It means that the size of a crumple of every scale is in the order of its size in the dense packing state and the crumples are mutually segregated in space. A more detailed elucidation of this problem has been presented in Ref. [65].

The system of densely packed globulized crumples corresponds to the chain with the fractal dimension $d_f = 3$ ($d_f = 3$ is realized from the minimal scale, g^*, up to the full globule size). The value g^* is in the order of

$$g^* = N_e(na^3)^{-2} \tag{50}$$

where N_e is the well-known parameter of reptation model and n is the globule density.

The last relation was obtained in Ref. [65] using the following arguments: $g = (na^3)^{-2}$ is the mean length of the chain part between two neighboring along the chain contacts with other parts; consequently $N_e g$ is the mean length of the chain part between topological contacts (entanglements). Of course, for phantom chains, Gaussian blobs of size g strongly overlap one another because pair contacts between monomers are insignificant under θ-conditions. However, for non-phantom chains, these pair contacts are essential from the topological point of view because chain crossings are prohibited independently of the value and sign of the virial coefficient.

The entropy loss connected with the crumpled state formation can be estimated as follows:

$$S \simeq -\frac{N}{g^*} \tag{51}$$

which together with Eq. (50) leads to the above estimate (45). Using Eq. (51), the corresponding crumpled globule density, n, can be obtained in the mean-field approximation by means of the minimization of its free energy. The density of the crumpled state is less than that of the usual equilibrium which is connected with additional topological repulsive-type interactions between crumples:

$$n_{crump} = \frac{n_{eq}}{1 + const(a^6/CN_e)} < n_{eq} \tag{52}$$

The estimate for the non-trivial knot formation probability, $p(\alpha)$ in the region $\alpha < 0.6$ with numerical constants obtained by comparing the relation (46) with numerical data of Ref. [29] is shown in Fig. 8 by the solid line for $N_e = 60$.

5 Conclusion

In conclusion, we would like to state our hypothesis concerning the possibility of reformulation of some topological problems for strongly collapsed chains in terms of integration over the set of trajectories with a fixed fractal dimension but without any topological constraints.

- We have argued above that in ensemble of strongly contracted unknotted chains (paths) most of them have the fractal dimension $d_f = 3$.
- We also believe that almost all paths in the ensemble of lines with fractal dimension $d_f = 3$ are topologically isomorphic with a trivial knot or at least with a sufficiently simple one.

Let us remind that the problem of the partition function calculation for a closed polymer chain with topological constraints is usually formulated as integration over the set Ω of closed paths from a fixed topological class, or with fixed value of the topological invariant:

$$Z = \int\int_\Omega D_w\{r\} \exp(-H) = \int \ldots \int D_w\{r\} \exp(-H)\delta(I - I_0) \tag{54}$$

where $D_w\{r\}$ means the integration with the usual Wiener measure and $\delta(I - I_0)$ cuts the paths with a fixed value of the topological invariant, I_0.

If our hypothesis is true, the problem for sufficiently contracted chains can be reduced to integration over all paths without any constraints, but with a special new measure, $\mathscr{D}_f\{r\}$:

$$Z = \int \ldots \int \mathscr{D}_f\{r\} \exp(-H) \tag{55}$$

The usual Wiener measure, $D_w\{r\}$, is concentrated on trajectories with fractal dimension $d_f = 2$. Instead of that, for description of an unknotted ring the measure $\mathscr{D}_f\{r\}$ with fractal dimension $d_f = 3$ should be used.

6 References

1. Gates V, et al. (1985) Physica D 15: 289
2. Witten E (1989) Commun Math Phys 121: 353; Frelich J, King C, (1989) Commun Math Phys 126: 167
3. Kholodenko AL (1991) Private communication
4. Nechaev S (1990) Int J Mod Phys (B), 4: 1809
5. de Gennes PG (1979) Scaling concepts in polymer physics, Cornell Univ Press, Ithaca, New York
6. Crowell RH, Fox RH (1963) Introduction to knot theory, Ginn, Boston
7. Vologodskii AV, Anshelevich VV, Lukashin AV, Frank-Kamenetskii MD (1984) Nature, 280: 294.
8. Vologodskii AV (private communication)
9. Grosberg A Yu, Zhestkov AV (1986) J Biomol Struct Dyn 3: 859
10. Edwards SF (1967) Proc Phys Soc 91: 513
11. Prager S, Frisch HL (1967) J Chem Phys 46: 1475
12. Saito N, Chen Y (1973) J Chem Phys 59: 3701
13. Grosberg A Yu, Khokhlov AR (1989) Statistical physics of macromolecules, Nauka, Moscow (in Russian)
14. Mackenzie R, Wilczek F (1988) I Mod Phys A 3: 2827
15. Tanaka F (1984) J Phys Soc Japan, 53: 2205
16. Helfand E, Pearson DS (1983) J Chem Phys 79: 2054
17. Rubinstein M, Helfand E (1985) J Chem Phys 82: 2477
18. Khokhlov AR, Nechaev SK (1985) Phys Lett 112-A: 156
19. Ternovskii FF, Khokhlov AR (1986) Zh Exp Teor Fiz 90: 1249
20. Cates ME, Deutsch JM (1986) J Physique 47: 2121
21. Rubinstein M (1986) Phys Rev Lett 57: 3023
22. Nechaev SK, Semenov AN, Koleva MK (1987) Physica 140-A: 506
23. Khokhlov AR, Ternovskii FF, Zheligovskaya EA (1990) Physica A 163: 747
24. Nechaev SK (1988) J Phys A Math Gen 21: 3659
25. Nechaev SK (1989) Europhys Lett 10: 317
26. Nechaev SK, Khokhlov AR (1988) Phys Lett 126-A: 431
27. Gerzenstein ME, Vasiljev VB (1959) Prob Theor Appl 4: 424; Karpelevich FI, Tutubalin VN, Shour MG (1959) Prob Theor Appl 4: 432 (in Russian)
28. Burke WL (1980) Spacetime geometry cosmology, Mill Valley, California
29. Vologodskii AV, Frank-Kamenetskii MD (1981) Usp Fiz Nauk 134: 641 (in Russian)
30. Vologodskii AV, Lukashin AV, Frank-Kamenetskii MD, Anshelevich VV (1974) Zh Exp Teor Fiz 66: 2153 (in Russian)
31. Vologodskii AV, Lukashin AV, Frank-Kamenetskii MD (1974) Zh Exp Teor Fiz 67: 1875 (in Russian)
32. Frank-Kamenetskii MD, Lukashin AV, Vologodskii AV (1975) Nature 258: 398
33. Klenin KV, Vologodskii AV, Anshelevich VV, Dykhne AM (1988) J Biomol Struct Dyn 5: 1173
34. Bret M le (1980) Biopolymers 19: 619
35. Mishels JP, Wiegel FW (1986) Proc Roy Soc (A) 403: 269
36. Koniaris K, Muthukumar M (1991) Phys Rev Lett 66: 2211
37. van Rensburg EJ, Whittington SG (1990) J Phys (A): Math Gen 23: 3573
38. Jones VFR (1985) Bull Am Math Soc 12: 103
39. Reidemeister K (1932) Knotentheorie Springer, Berlin
40. Kauffman LH (1989) Braid Group, Knot theory and statistical mechanics, World Scientific, Singapore
41. Kauffman LH (1987) Topology 26: 395
42. Wadati M, Degushi T, Akutsu Y (1989) Phys Rep 180: 247
43. Baxter RJ (1982) Exactly solvable models in statistical mechanics, Academic, NY
44. Turaev VG (1987) LOMI-preprint E-3-87
45. Jones VFR (1989) Pacific J Math 137: 311
46. Lickorish WBR (1988) Bull London Math Soc 20: 558
47. Grosberg A, Nechaev S (1992) J Phys (A): 25: 4659
48. Witten E (1989) Nucl Phys (B) 322: 629

49. Guadarini E, Martinelli M, Mintchev M (1989) CERN-TH preprints 5419/89, 5420/89, 5479/89
50. de Gennes PG (1971) J Chem Phys 55: 572
51. Doi M, Edwards SF (1978) J Chem Soc Faraday 2, 74: 1789, 1802, 1819; Doi M, Edwards SF, The Theory of Polymer dynamics, Clarendon, Oxford, 1986
52. Heinrich G, Helmis G, Straube E (1988) Adv Polym Sci 85: 33
53. Evans KE, Edwards SF (1981) J Chem Soc Faraday 2, 77: 1891, 1913, 1929; Baumgartner A, Muthukumar M (1986) J Chem Phys 84: 440
54. Pakula T, Geyler S (1988) Macromolecules 21: 1665
55. Lifshits IM, Grosberg A Yu (1973) Zh Exp Teor Fiz 65: 2399
56. Semenov AN (1986) Zh Exp Teor Fiz 91: 122 (in Russian)
57. Lomakin AV, Semenov AN (1988) Zh Exp Teor Fiz 94: 138 (in Russian)
58. Semenov AN (1988) J Physique 49: 175
59. Semenov AN (1988) J Physique 49: 1353
60. Panykov SV (1988) Zh Exp Teor Fiz 94: 174 (in Russian)
61. Goldbart P, Goldenfeld N (1989) Phys Rev 39-A: 1402, 1412
62. Chowdhury D (1986) Spin glasses and other frustrated systems, World Scientific, Singapore
63. Mezard M, Parisi G, Virasoro M (1987) Spin glasses theory and beyond, World Scientific, Singapore
64. Grosberg A Yu, Nechaev SK, Shakhnovich EI (1988) Biofizika 33: 247 (in Russian)
65. Grosberg A Yu, Nechaev SK, Shakhnovich EI (1988) J Physique 49: 2095
66. Fixmann M (1962) J Chem Phys 36: 306
67. Grosberg A Yu, Nechaev SK (1991) Macromolecules 24: 2789

Editor: K. Dušek
Received January 1992

Phase Behavior of Polymer Blends – Effects of Thermodynamics and Rheology

H.W. Kammer, J. Kressler, and C. Kummerloewe
Department of Chemistry, Dresden University of Technology,
Mommsenstraße 13, 0-8027 Dresden, FRG

This article reviews the phase behavior of polymer blends with special emphasis on blends of random copolymers. Thermodynamic issues are considered and then experimental results on miscibility and phase separation are summarized. Section 3 deals with characteristic features of both the liquid–liquid phase separation process and the reverse phenomenon of phase dissolution in blends. This also involves morphology control by definite phase decomposition. In Sect. 4 attention will be focused on flow-induced phase changes in polymer blends. Experimental results and theoretical approaches are outlined.

1 Introduction

Recently, there has been pronounced interest in polymer blends. The enhanced activities are related to both the hope of producing advanced high-performance materials based on well-known products and the need of basic knowledge on their phase behavior which in turn offers an opportunity for morphology control during processing. Polymer blends are combinations of at least two polymer components that can either mix completely on a molecular scale or, what is more often observed, form a heterogeneous, two-phase mixture. Therefore, with respect to their phase behavior, polymer blends can be characterized as being either miscible or immiscible. The term "miscibility of polymers" will be used for their dispersal at the molecular level.

Advances in Polymer Science, Vol. 106
© Springer-Verlag Berlin Heidelberg 1993

The phase behavior of polymer blends comprising amorphous polymers is experimentally well accessible in a "window" which is bounded at high temperatures by the thermal decomposition temperature of the polymer components and at low temperatures by the glass transition temperature of the system (cf. Fig. 1). Below the glass transition temperature the phase behavior can be estimated only tentatively.

To avoid confusion, one has to be conscious of the method used to judge the phase situation of a blend. Perhaps the most commonly used and widely accepted criteria for miscibility are optical clarity and the existence of a single glass transition temperature for a miscible blend. However, one has to recognize that transparency is only indicative of miscibility when the refractive indices of the two components are sufficiently dissimilar (> 0.01). Furthermore, both methods are sensitive to heterogeneities of more than approximately 50 nm in domain size. Thus, if a miscible blend contains heterogeneous regions, these are on a submolecular scale. For a more refined assurance of miscibility, i.e. miscibility at the segmental level, methods such as NMR or excimer fluorescence should be employed.

Historically, it was believed that miscibility of high-molar-mass polymers at a molecular level can rarely be observed owing to the extremely small combina-

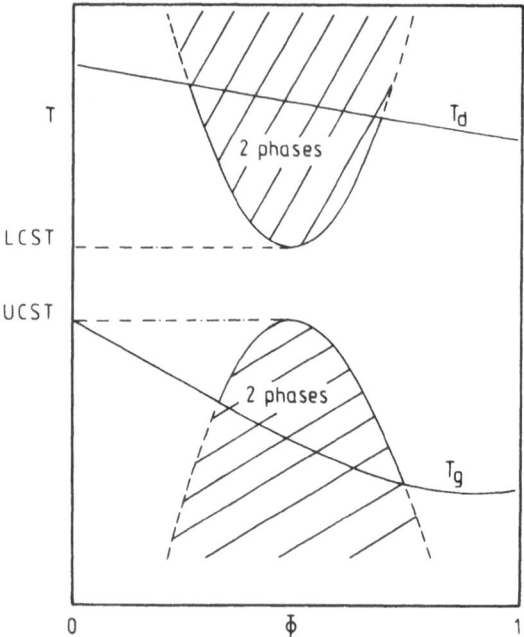

Fig. 1. True equilibrium behavior in polymer blends can be detected in the window bounded by the thermal decomposition temperature (T_d) and the glass-transition temperature (T_g)

torial entropy of mixing. In comparison to low-molar compounds the number of molecules in a volume unit is reduced by a factor of $1/r$ (r being the degree of polymerization) which is accompanied by a correspondingly diminished entropy of mixing. Therefore, for homopolymers to be miscible the existence of specific interactions is necessitated leading to a negative enthalpy of mixing. However, intensive research over the last few years has revealed that miscibility in polymer systems containing random copolymers at least as one component is not nearly as rare as once thought. Miscibility of homopolymers and random copolymers is affected by the so-called repulsion effect [1-3]. It must be emphasized that for these systems the entropy of mixing and the enthalpy of mixing are of the same order of magnitude [4-7]. Therefore, the equilibrium phase behavior of blends based on copolymers is ruled by a subtle balance of small energy and entropy effects which may not be significant in small molecule systems and which causes rather unusual phase diagrams. Intra- as well as intermolecular interactions play an important role in determining the phase behavior.

In general, miscible blends display phase separation at elevated temperatures, as shown schematically in Fig. 1, i.e. lower critical solution temperature (LCST) behavior can be seen. As a general phenomenon, miscibility of polymers must be coupled with ordering in the system imposed by specific interactions or by the repulsion effect. An increase in temperature weakens the specific interactions and the repulsion effect, respectively, which is equivalent to ascending disorder. Hence, an entropy-driven LCST occurs where the blend phase-separates upon heating. Some miscible blends exhibit not only LCST behavior but also thermally induced phase separation upon cooling. Cooling corresponds to a decrease in compressibility which in turn is equivalent to enhanced repulsion between the segments exceeding the specific interactions below an upper critical solution temperature (UCST). Thus, the repulsion between the segments turns out to be unfavorable for order or miscibility below a certain temperature and promotes phase separation. The simultaneous occurrence of an LCST as well as a UCST in blends of high-molar-mass polymers is considered to be a general phenomenon [8]. But usually, the UCST shifts far below the glass transition temperature and, therefore, is not accessible experimentally. When the glass transition temperature is sufficiently low as in systems containing an elastomer as one of the components the UCST could be confirmed experimentally besides an LCST [9-12].

The dynamic phase behavior of polymer blends is also unusual. The high viscosity slows down the diffusion processes to the point where the early stages of the spinodal decomposition can be monitored experimentally [13]. Many experimental and theoretical studies have been made on the kinetics of phase separation in initially homogeneous mixtures set isothermally into the thermodynamically unstable region [14-18]. The unmixing in the unstable region proceeds via spinodal decomposition. This means for the early stage of phase separation that the growth of the thermally induced concentration fluctuations is caused entirely by a thermodynamic driving force which initiates a diffusional flux against the concentration gradient.

Various polymer blends display, just above the LCST and also below the UCST, a very regular, network-like two-phase morphology. This phenomenon conveys the possibility of manipulating and controlling the phase morphology in polymer blends [19–21].

Finally, a challenging problem is to discuss the influence of hydrodynamic flow fields on the phase behavior of polymer blends. This is of fundamental interest and of technological importance as well since stresses and corresponding deformations are encountered during processing of blends. Extension of studies to blend systems under external flow is necessary for the better understanding of structure formation in polymer blends outside equilibrium.

This review has three general aims:

- The first is to develop thermodynamic issues to understand the complex phase behavior of polymer blends. Experimental determination of miscibility regions provides the individual segmental interaction parameters necessary for predictions of various phase equilibria.
- The second is to examine the dynamics of phase separation and phase dissolution which can be pursued by scattering techniques. This topic involves the fundamental problem of self-organization in polymer systems under non-equilibrium conditions.
- Finally, the third purpose is to outline the influence of flow on the phase behavior of blends. This will include both experimental results and theoretical approaches to reveal the effect of different flow conditions on the phase behavior.

2 Equilibrium Phase Behavior

2.1 Thermodynamics of Polymer–Polymer Miscibility

According to the Second Law of thermodynamics the phase behavior of polymer mixtures, like the behavior of any other mixture, is controlled by two thermodynamic factors, the entropy of mixing which consists of combinatorial and noncombinatorial contributions and the enthalpy of mixing which is related to the interactions between the segments. One may establish that the combinatorial part of the entropy of mixing generally supports miscibility whereas the enthalpy favors miscibility only for exothermic interaction effects. The noncombinatorial entropy of mixing and the enthalpy of mixing represent the excess free enthalpy of mixing, ΔG_E^M. Therefore, the Gibbs free energy of mixing can be expressed as

$$\Delta G^M = \Delta G_E^M - T \cdot \Delta S_{comb}^M \qquad (1)$$

where ΔS_{comb}^M is the combinatorial entropy of mixing which can be described in the context of the Flory–Huggins theory for a binary mixture (ϕ-volume fraction) by

$$-\frac{\Delta S_{comb}^M}{R} = \frac{\phi_A}{r_A} \ln \phi_A + \frac{\phi_B}{r_B} \ln \phi_B \qquad (2)$$

where r_i is the degree of polymerization of component i. The quantity ΔG_E^M, on the other hand may be represented by

$$\frac{\Delta G_E^M}{RT} = \phi_A \phi_B X \qquad (3)$$

As can be seen, X is a dimensionless free-energy parameter which is sometimes called a generalized interaction parameter.

The conditions for phase stability in a binary mixture of composition ϕ at fixed temperature T and pressure P are

$$\Delta G^M < 0 \qquad \left(\frac{\partial^2 \Delta G^M}{\partial \phi^2}\right)_{P,T} > 0 \qquad (4)$$

Employing Eqs. (2) and (3) and assuming the parameter X as independent of concentration one gets

$$\left(\frac{\partial^2 \Delta G^M}{\partial \phi^2}\right)_{P,T} = \left(\frac{1}{r_A \phi_A} + \frac{1}{r_B \phi_B} - 2X\right) RT \qquad (5)$$

It emanates from Eq. (5) that as mentioned before the combinatorial entropy of mixing stabilizes the mixture and that $X < 0$ favors miscibility of the components, especially, in the case of high-molar-mass polymers (each r large) when the combinatorial entropy of mixing tends to very small values.

To discuss the phase stability of polymer blends in more detail one has to specify the free-energy parameter X. This can be done in terms of an equation-of-state theory [8]. Theories that take into account the compressible nature of the pure components as well as that of the mixture are called equation-of-state theories. As basic quantities characterizing the thermodynamic state of a system the reduced temperature ($\tilde{T}$), volume ($\tilde{V}$) and pressure ($\tilde{P}$) are employed and defined by

$$\tilde{T} \equiv T/T^* \qquad \tilde{V} \equiv V/V^* \qquad \tilde{P} \equiv P/P^* \qquad (6)$$

where the starred quantities are reference parameters.

Expressing the potential energy $\varepsilon(r')$ of a pair of segments at the distance r', which belongs to r-mers, by the general function ζ with two scale factors ε^* and r^* characteristic of the molecular species:

$$\varepsilon(r') = -\varepsilon^* \zeta(r'/r^*) \tag{7}$$

then one may express, in the simplest approximation, the starred quantities in Eq. (6) by the molecular parameters as follows

$$T^* = \varepsilon^*/k \qquad V^* = r^{*3} \qquad P^* = \varepsilon^*/V^* \tag{7'}$$

The scale factors ε^* and r^* represent the coordinates of the minimum of the potential function $\varepsilon(r')$. – In generalization of Eq. (7) for a mixture of two homopolymers A and B consisting of segments of types A and B one may introduce the average interaction of a segment (say A) with the neighbor segments at a distance r' in terms of the volume fraction ϕ

$$\langle \varepsilon_A^* \rangle \zeta(r'/\langle r_A^* \rangle) = \phi_A \varepsilon_{AA}^* \zeta(r'/r_{AA}^*) + \phi_B \varepsilon_{AB}^* \zeta(r'/r_{AB}^*) \tag{8}$$

As can be seen, the interaction energy $\langle \varepsilon_A(r') \rangle$ is supposed to be of the same form as in Eq. (7), however, the scale factors $\langle \varepsilon_A^* \rangle$ and $\langle r_A^* \rangle$ are now average composition-dependent parameters. Moreover, for blends based on random copolymers the quantities ε_{AA}^*, r_{AA}^* et cetera of the right-hand side of Eq. (8) also have to be replaced by average parameters $\langle \varepsilon_{AA}^* \rangle$, $\langle r_{AA}^* \rangle$ and so on. The latter mean values are now number averages over the copolymer composition in terms of mole fractions.

For a mixture of two homopolymers consisting of segments of types A and B, respectively, there exist six scale factors ε_{ij}^*, r_{ij}^* $(i, j = A, B)$. Choosing one component (say A) as the reference substance one may use them to define four dimensionless parameters characteristic for the mixture:

$$X_{AB} \equiv \frac{1}{\varepsilon_{AA}^*} \left[\tfrac{1}{2}(\varepsilon_{AA}^* + \varepsilon_{BB}^*) - \varepsilon_{AB}^* \right] \qquad R_{AB} \equiv \frac{1}{r_{AA}^*} \left[\tfrac{1}{2}(r_{AA}^* + r_{BB}^*) - r_{AB}^* \right] \tag{9}$$

$$\Gamma \equiv \frac{\varepsilon_{BB}^*}{\varepsilon_{AA}^*} - 1 \qquad\qquad \rho \equiv \frac{r_{BB}^*}{r_{AA}^*} - 1$$

For blends containing random copolymers the parameters are more complicated owing to the fact that the scale factors must be replaced by average quantities $\langle \varepsilon_{AA}^* \rangle$ and so on. Explicit expressions are given in Appendix I.

Using (8) and (7') it follows immediately that the reduced quantities of (6) generally must be recast in average composition-dependent variables

$$\langle \tilde{Y} \rangle \equiv Y/\langle Y^* \rangle \qquad (Y = T, V, P) \tag{6'}$$

For ordinary pressures, to a good approximation, it is justified to assume $\tilde{P} = 0$. As a result, the parameter X is a function of the reduced temperature and volume

$$X = X(\langle \tilde{T} \rangle, \langle \tilde{V} \rangle) \tag{10}$$

Remembering that according to Eq. (3) and in the limit $\Delta G^M = \Delta F^M$

$$X\phi_A\phi_B = \phi_A(\bar{F}_A - F_A) + \phi_B(\bar{F}_B - F_A) - \phi_B(F_B - F_A) \tag{11}$$

holds good where $\bar{F}_i$ and F_i are the partial molar and the molar free energy of component i per RT, respectively, one may expand Eq. (11) in powers of $1/\langle \tilde{T} \rangle$ and $\langle \tilde{V} \rangle$ around component A. For the sake of simplicity we assume that component A is a homopolymer then the average quantity $\langle \tilde{T}_{AA} \rangle$ simplifies to $\tilde{T}_{AA}$ and will be symbolized simply by $\tilde{T}_A$ and analogously the quantity $\langle \tilde{V}_{AA} \rangle$ is replaced by $\tilde{V}_A$. It follows when higher than second order derivatives are ignored

$$X\phi_A\phi_B = \left(\frac{\partial X}{\partial \frac{1}{\tilde{T}}}\right)_A (\phi_A\theta_A + \phi_B\theta_B - \phi_B\theta_{BB})\frac{1}{\tilde{T}_A}$$

$$+ \frac{1}{2}\left(\frac{\partial^2 X}{\left(\partial \frac{1}{\tilde{T}}\right)^2}\right)_A (\phi_A\theta_A^2 + \phi_B\theta_B^2 - \phi_B\theta_{BB}^2)\frac{1}{\tilde{T}_A^2}$$

$$+ \left(\frac{\partial^2 X}{\partial \frac{1}{\tilde{T}}\partial\tilde{V}}\right)_A (\phi_A\theta_A\Omega_A + \phi_B\theta_B\Omega_B - \phi_B\Omega_{BB}\theta_{BB})\frac{1}{\tilde{T}_A}$$

$$+ \frac{1}{2}\left(\frac{\partial^2 X}{\partial\tilde{V}^2}\right)_A (\phi_A\Omega_A^2 + \phi_B\Omega_B^2 - \phi_B\Omega_{BB}^2) \tag{12}$$

where

$$\theta_i \equiv \frac{\tilde{T}_A}{\langle\tilde{T}_i\rangle} - 1 = \frac{\langle\varepsilon_i^*\rangle}{\varepsilon_{AA}^*} - 1 \qquad \theta_{BB} \equiv \frac{\tilde{T}_A}{\langle\tilde{T}_{BB}\rangle} - 1 = \frac{\langle\varepsilon_{BB}^*\rangle}{\varepsilon_{AA}^*} - 1$$

$$\Omega_i \equiv \langle\tilde{V}_i\rangle - \tilde{V}_A \qquad\qquad \Omega_{BB} \equiv \langle\tilde{V}_{BB}\rangle - \tilde{V}_A \quad (i = A, B) \tag{13}$$

From $X = \Delta F^M/(RT\phi_A\phi_B)$ the derivatives occurring in Eq. (12) can be easily calculated using thermodynamic standard relations (cf. Appendix II). Furthermore, the quantities θ_i and Ω_i can be expressed by the parameters defined in Eq. (9) as given in Appendix III. Finally, inserting the respective expressions of Appendices II and III into Eq. (11) and neglecting second-order terms in X_{AB} and R_{AB} and also second-order combinations of them with the parameters Γ and ρ, one arrives at

$$X = -\frac{U_A}{RT}\left(2X_{AB}^T + \frac{\tilde{V}_A^2}{\tilde{\kappa}_A}\frac{9}{8}\rho^2\right) + \frac{C_{VA}}{R}\frac{7}{8}\Gamma^2 \tag{14}$$

where

$$X_{AB}^T \equiv X_{AB} + \frac{9}{2}\rho^2 - \frac{3}{8}\Gamma\rho$$

According to Eq. (14) three effects contribute to the excess free enthalpy of mixing:

1) the segmental interaction represented by the parameter X_{AB};

2) the free-volume effect arising from the difference in the thermal expansion coefficients of the components and symbolized by the parameter Γ;

3) the size-effect resulting from the difference in the sizes of the segments and expressed by the parameter ρ.

The prefactors in the three terms of Eq. (14) regulate the temperature dependence of the previously mentioned effects. The molar configurational energy $-U_A$, the heat capacity C_{VA} and the reduced compressibility $\tilde{\kappa}_A$ are positive definite. Moreover, the molar configurational energy $-U_A$ and the quantity $-U_A\tilde{V}_A^2/\tilde{\kappa}_A$ are decreasing functions of temperature while the heat capacity ascends with increasing temperature. As a result, the interaction and the size-effect decrease with increasing temperature whereas the free-volume effect increases with temperature. Equation (14) indicates that the requirement $X < 0$ for miscibility can only be fulfilled when the interaction parameter X_{AB} is negative. For $X_{AB} < 0$, the interaction term of Eq. (14) is also negative and favors mixing.

The temperature dependence of the prefactors of Eq. (14) can be shown explicitely when one replaces the quantities $-U_A$, C_{VA} and $\tilde{\kappa}_A$ by the reduced volume $\tilde{V}_A$ applying a suitable equation-of-state. According to Flory [22] the following equation links the reduced thermodynamic variables

$$\tilde{P} = \frac{\tilde{T}\tilde{V}^{-2/3}}{\tilde{V}^{1/3} - 1} - \frac{1}{\tilde{V}^2} \tag{15}$$

which simplifies in the limit $\tilde{P} = 0$ to

$$\tilde{T} = \frac{\tilde{V}^{1/3} - 1}{\tilde{V}^{4/3}} \tag{15'}$$

$\tilde{V}^{1/3}$ varies in the range $1 \ldots 4/3$. From Eq. (15') one immediately obtains a relation between the reduced volume $\tilde{V}$ and the thermal expansion coefficient α:

$$\tilde{V}^{1/3} - 1 = \frac{\alpha T}{3(1 + \alpha T)} \tag{16}$$

From a single temperature where α is known the reference temperature T_A^* can be calculated from Eqs. (15') and (16) for the component A which is taken as the reference substance.

Employing Eqs. (15') and (15) one can easily calculate the expressions for the prefactors given in Appendix IV. These relations combined with Eq. (14) show that the empirical representation of the interaction parameter by

$$X = \frac{A}{T} + B \tag{17}$$

is approximately justified. However, as Eq. (14) also shows, the quantities A and B are not constants but depend on temperature.

The parameter X of Eq. (14) and its constituents as a function of temperature are depicted in Fig. 2. For $X_{AB} < 0$, the interaction term of Eq. (14) is also negative and favors mixing. In the case $\rho = 0$, it dominates the unfavorable free-volume term up to a certain temperature, i.e. miscibility can be observed at

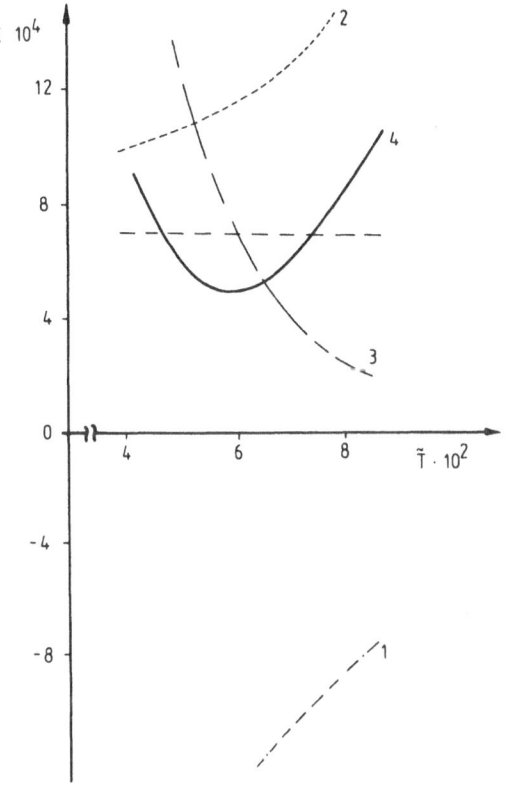

Fig. 2. Variation of the parameter X (*curve 4*) and its constituents – interaction (*1*), free-volume (*2*) and size-effect (*3*) – as a function of reduced temperature according to Eq. (14). The parameters used are: $X_{AB}^T = -1 \times 10^{-4}$, $\Gamma^2 = 6 \times 10^{-4}$, $\rho^2 = 3 \times 10^{-5}$. The combinatorial entropy of mixing at $\phi = 0.5$ and $r = 1000$ is given by the horizontal dashed straight line

low temperatures. With increasing temperature the magnitude of the interaction term decreases and the free-volume term increases leading to an increase of the parameter X. As a result, a LCST occurs followed by phase separation at still higher temperatures. This is the usual phase behavior of polymer blends regulated by the interplay of interaction and free-volume effects. When the parameter $\rho \neq 0$, the second term of Eq. (14) increases with decreasing temperature and exceeds the interaction term below a certain temperature whereas it is negligible at sufficiently high temperatures. Therefore, the interaction term dominates the unfavorable free-volume and size-effect terms only within a certain range of temperatures, i.e. the system displays not only a LCST but also a UCST.

In the limit the degree of polymerization r tends to infinity, UCST and LCST are given by X = 0. From Eq. (14) combined with the expressions of Appendix

IV the locations of the critical points can be obtained:

$$\text{UCST:} \quad \tilde{V}_A^{1/3} = 1 + \frac{\rho^2}{-\frac{16}{3} X_{AB}^T + 3\rho^2 + \frac{7}{3}\Gamma^2} \tag{18}$$

$$\text{LCST:} \quad \tilde{V}_A^{1/3} = \frac{4}{3} - \frac{\frac{7}{9}\Gamma^2}{-\frac{16}{3} X_{AB}^T + 3\rho^2 + \frac{7}{3}\Gamma^2} \tag{19}$$

It follows that the positions of the UCST and LCST are chiefly determined by the ratios $\rho^2/|X_{AB}^T|$ and $\Gamma^2/|X_{AB}^T|$, respectively.

As mentioned before, Eq. (3) reveals that the parameter X is a free-energy parameter comprising an enthalpic part X_H and an entropic part X_S:

$$X = X_H + X_S \qquad X_S = \frac{\partial}{\partial T}(TX) \tag{20}$$

Notice that $X_H \sim \Delta H^M$ and $X_S \sim -\Delta S_E^M$. Applying Eq. (14) one gets

$$X_S = -\frac{C_{VA}}{R} 2X_{AB}^T + \left[\frac{1}{\tilde{T}_A \tilde{V}_A} - \frac{\tilde{V}_A}{\tilde{\kappa}_A^2}\frac{\partial \tilde{\kappa}_A}{\partial \tilde{T}_A}\right] \cdot \frac{9}{8}\rho^2$$
$$+ \frac{C_{VA}}{R}\left[1 + \frac{4}{9}\tilde{T}_A \tilde{V}_A^{2/3}\left(\frac{C_{VA}}{R}\right)^2\right] \cdot \frac{7}{8}\Gamma^2 \tag{21}$$

For $X_{AB}^T < 0$ the only negative contribution to X_S results from the third term on the right-hand side which becomes significant only at sufficiently low temperatures. Therefore, miscibility in polymers is accompanied by a negative non-combinatorial entropy effect or, in other words, by ordering imposed by specific interactions. At sufficiently high temperatures this negative non-combinatorial entropy contribution counterbalances the favorable energetic effect and induces phase separation. It becomes obvious that in phase separation at elevated temperatures, the LCST, turns out to be entropy driven. In contrast to the LCST, the phase separation at low temperatures is caused by the enthalpic contribution to the parameter X associated with the size-effect. Therefore, order in miscible polymers is disorganized at low temperatures by an additional repulsion. The variation of the enthalpic and entropic contribution to X with respect to reduced temperature is depicted in Fig. 3.

Volume of Mixing. In general terms, exothermic interaction effects tend to diminish the volume of a mixture whereas entropic effects act in the opposite way. For miscible polymers, therefore, one expects a negative volume of mixing. This has been confirmed experimentally for different miscible polymers with LCST behavior, e.g. for miscible 50/50 blends of polystyrene and poly(2-chlorostyrene) the volume change $\Delta V^M/V$ at 130 °C has been reported to be about

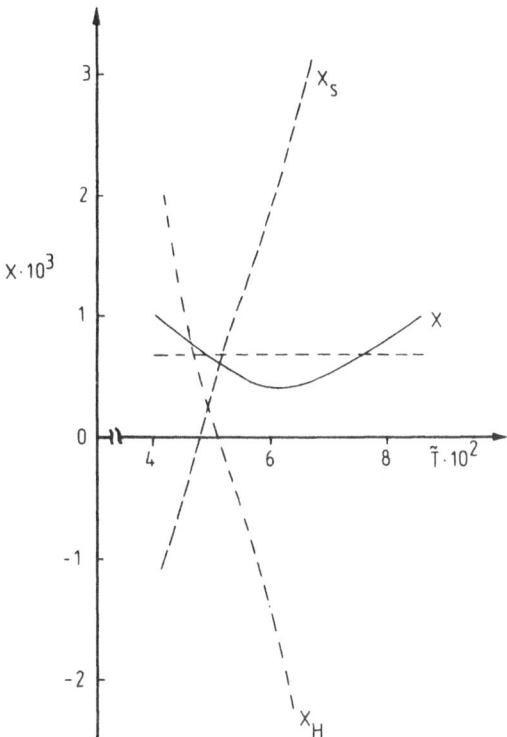

Fig. 3. Representation of the parameter X and its enthalpic and entropic parts, X_H and X_S, respectively, versus reduced temperature as calculated from Eqs. (20) and (21). The parameters are the same as in Fig. 2

-1×10^{-2} [23]. However, when the size-effect comes into play or, in other words, the blend displays simultaneously LCST and UCST then also positive volumes of mixing may occur. This effect has been observed in blends of poly(methyl methacrylate) (PMMA) and poly(ethyl acrylate) with poly(vinylidene fluoride) (PVDF) [24]. Positive volumes of mixing can be explained in the context of the equation-of-state theory reported above. When one calculates the volume of mixing in the same approximation as the Gibbs free energy parameter X (cf. Eq. (14)) it follows that [25]:

$$\frac{\Delta \tilde{V}^M}{\tilde{V}_A \phi_A \phi_B} = \frac{3}{4} \rho \left(\Gamma + \frac{11}{2} \rho \right) + \tilde{\alpha}_A \tilde{T}_A \left(2 X_{AB} - \frac{3}{4} \Gamma^2 + 9 \rho^2 + \frac{9}{4} \rho \Gamma \right)$$

$$- \tilde{T}_A^2 \cdot \left(\tilde{\alpha}_A^2 + \frac{\partial \tilde{\alpha}_A}{\partial \tilde{T}_A} \right) \cdot \frac{3}{8} \Gamma^2 \qquad (22)$$

where $\tilde{\alpha}_A$ is the reduced thermal expansion coefficient. As can be seen from Eq. (22), when there is no size effect in miscible polymers, i.e. $\rho = 0$ and $X_{AB} < 0$, then the volume of mixing ΔV^M is negative. With increasing ρ the sign of ΔV^M will change whereas $\partial^2 \Delta G^M / \partial \phi^2$ keeps the same sign. As a result, $\Delta V^M > 0$ can occur for miscible systems.

The pressure dependence of the critical temperature can be extracted from thermodynamic standard relations to be

$$\left(\frac{\partial T}{\partial P}\right)_c = \frac{\Delta V_c^{M''}}{\Delta S_c^{M''}} \tag{23}$$

where the quantities of the right-hand side represent the second derivative of the volume of mixing and the entropy of mixing, respectively, with respect to volume fraction at the critical point. Keeping in mind that $\Delta S^{M''}$ is proportional to X_S one can easily extract from Fig. 3

$$\Delta S^{M''} \gtrless 0 \text{ at } \begin{array}{l} \text{LCST} \\ \text{UCST} \end{array} \tag{24}$$

For miscible blends with $\Delta V^M < 0$ near LCST and $\Delta V^M > 0$ in the vicinity of UCST one gets

$$\Delta V^{M''} \gtrless 0 \text{ at } \begin{array}{l} \text{LCST} \\ \text{UCST} \end{array} \tag{25}$$

Hence, both UCST and LCST shift to higher temperatures with increasing pressure. However, if $\Delta V^M > 0$ in the gap between UCST and LCST then the gap diminishes with increasing pressure. As a rule, taken from the theory, positive volumes of mixing are likely when the gap between LCST and UCST is sufficiently small. In other words, an increasingly positive volume of mixing is unfavorable for miscibility of polymers and leads ultimately to phase instability.

For the system PMMA/PVDF one can estimate the volume of mixing according to Eq. (22). As the key-point, the system exhibits both LCST and UCST. The critical points are reported to be about at 325 and 140 °C for 50/50 blends [11]. These data can be used to calculate, from Eqs. (18) and (19), the quantities X_{AB} and ρ.

The polymer PMMA is taken to be the reference substance A. The thermal expansion coefficients for PMMA and PVDF have been found to be $\alpha = 6.3 \times 10^{-4}\,K^{-1}$ [26] and $7.6 \times 10^{-4}\,K^{-1}$ [27], respectively, in the range 150 to 200 °C. Employing Eqs. (16) and (15') one immediately obtains the reference temperatures:

$$\text{PMMA: } T_A^* = 8100\,K \qquad \text{PVDF: } T_B^* = 7320\,K.$$

The quantity Γ follows from Eq. (9): $\Gamma^2 = 9.3 \times 10^{-3}$. Moreover, we note after applying of Eq. (15'): UCST (at 140 °C) and LCST (at 325 °C) correspond to $\tilde{V}^{1/3} = 1.0658$ and 1.1135, respectively. Thus, the quantities X_{AB} and ρ can be calculated from Eqs. (18) and (19). It follows that

$$X_{AB} = -9 \times 10^{-3} \qquad \rho^2 = 2.2 \times 10^{-3}$$

where X_{AB} agrees pretty well with the result $X_{AB} = -7 \times 10^{-3}$ submitted in [28]. – Using the values of the parameters presented previously, one can calculate the excess volume according to Eq. (22) for a 50/50 blend of PMMA and PVDF at 447 K ($\tilde{V}_A^{1/3} = 1.0732$). This gives

$$\frac{\Delta V^M}{V} = 3.2 \times 10^{-3}$$

in excellent agreement with the experimental result 2×10^{-3} [24].

In conclusion, the mean-field theory outlined above turns out to be a powerful tool for rationalizing the complex phase behavior of polymer blends, especially of random copolymer based blends, in terms of interaction, free-volume and size effects.

2.2 Phase Behavior of Blends with Random Copolymers

In the following, the phase behavior of copolymer-based blends will be illustrated by listing experimental results which have been published chiefly in the last five years. Phase behavior of homopolymer/copolymer and copolymer/copolymer blends can be rationalized in temperature-copolymer composition and isothermal copolymer composition – copolymer composition plots, respectively, at fixed blend ratio. The blend systems will be classified as follows:

1) homopolymer/random copolymer blends;
2) blends of two chemically identical random copolymers of the type $A_x B_{1-x}$, but differing in the composition x;
3) blends of two random copolymers having a common segment;
4) blends of two completely different random copolymers.

1) *Homopolymer (A)/random copolymer (B) blends, poly(1)/poly(2-ran-3)*. Phase behavior may be discussed in terms of the interaction parameter X_{AB} which is given in the mean-field approximation by [2, 3]

$$X_{AB} = x \chi_{12} + (1 - x) \chi_{13} - x(1 - x) \chi_{23} \qquad (26)$$

where x and χ_{ij} are the mole fraction of units 2 in the copolymer and the individual interaction parameters, respectively. The condition $X_{AB} < 0$, necessary for miscibility, leads to the following three cases

$- \chi_{12} < 0, \chi_{13}, \chi_{23} > 0$	miscibility door
$- \chi_{23} > \chi_{12}, \chi_{13} > 0$	window of miscibility
$- \chi_{ij} < 0$	window of immiscibility

Schematic representations of the respective miscibility regions, when only LCST behavior occurs, are shown in Fig. 4. As far as the authors are aware, a window of immiscibility has not yet been reported.

Miscibility doors can be observed when the homopolymer A is miscible with the homopolymer consisting of segments of type 2. Usually, only very near to the miscibility-immiscibility boundary can a temperature dependence of the phase behavior be seen, i.e. an LCST occurs. Figure 5 shows examples for miscibility doors. Further systems are listed in Table 1. Miscibility doors were also observed for blends of styrene copolymers and poly(vinyl methyl ether) (PVME) (Fig. 6, Table 2). In contrast to PPO/PS systems blends of PVME and PS

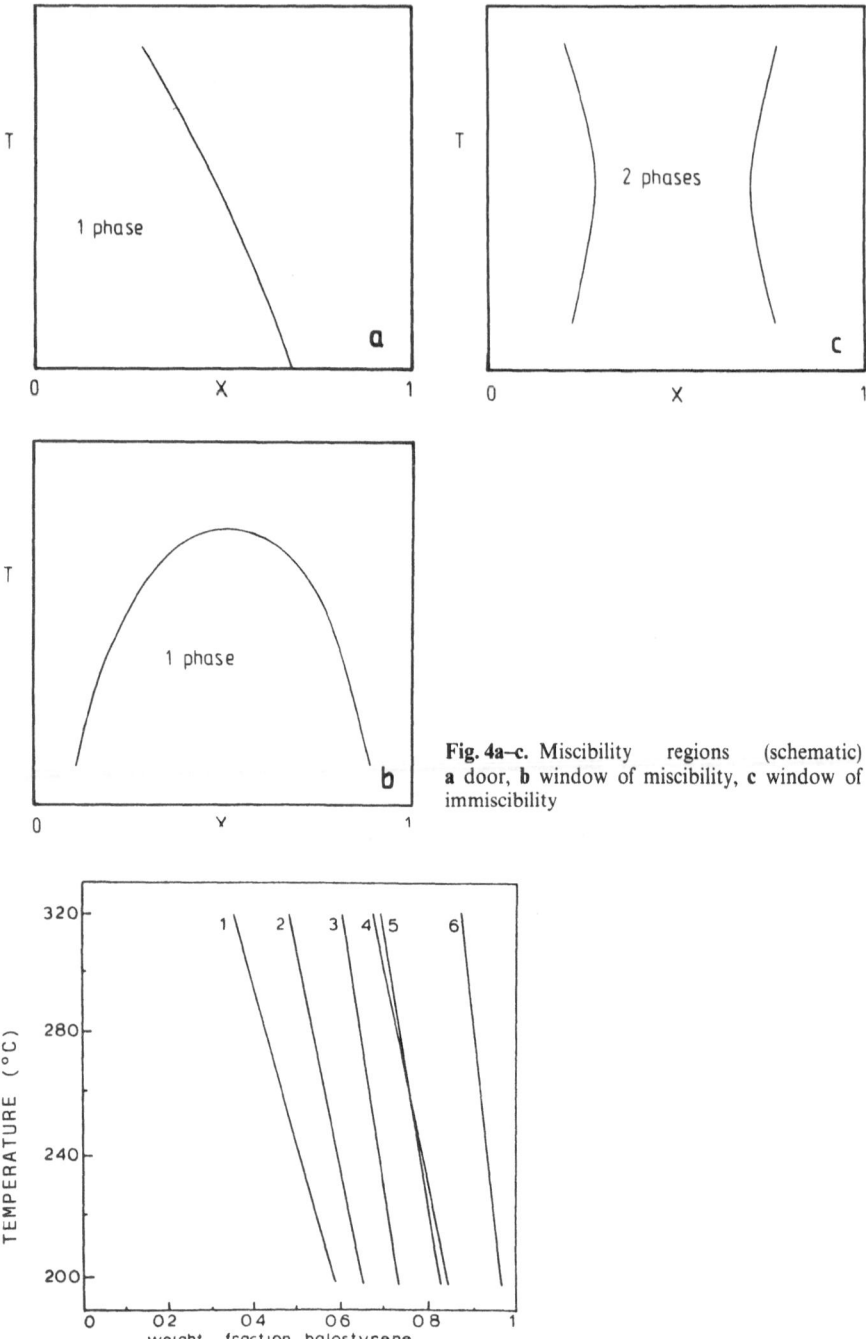

Fig. 4a–c. Miscibility regions (schematic) **a** door, **b** window of miscibility, **c** window of immiscibility

Fig. 5. Miscibility of poly(2,6-dimethyl phenylene oxide) (PPO) and random copolymers of styrene and *o*-bromostyrene (*1*), *p*-fluorostyrene (*2*), *p*-bromostyrene (*3*), *o*-chlorostyrene (*4*), *p*-chlorostyrene (*5*), and *o*-fluorostyrene (*6*) for 50/50 blends. Miscibility occurs to the left of the curves (after Ref. [29])

Table 1. Miscibility doors of PPO blended with styrene-based copolymers

Copolymer	Miscibility domain	Ref.
poly(styrene-*ran*-methyl methacrylate) (SMMA)	MMA content ≤ 29.2 wt% and more than 40% PPO in the blend	[30]
	MMA content ≤ 18 wt%, complete miscibility	[31]
poly(S-*ran*-acrylonitrile) (SAN)	AN content ≤ 10 wt%, complete miscibility	[32]
poly(S-*ran*-2,2,6,6-tetramethyl piperidinyl methacrylate) (STMPA)	TMPA content < 36 wt% and PPO content in the blend > 40%	[33]
poly(S-*ran*-iodinated S)	content of iodinated units ≤ 47 wt%; complete immiscibility ≥ 61 wt%	[34]
poly(*p*-methyl styrene-*ran*-AN)	AN content ≤ 7.7 wt%	[35]

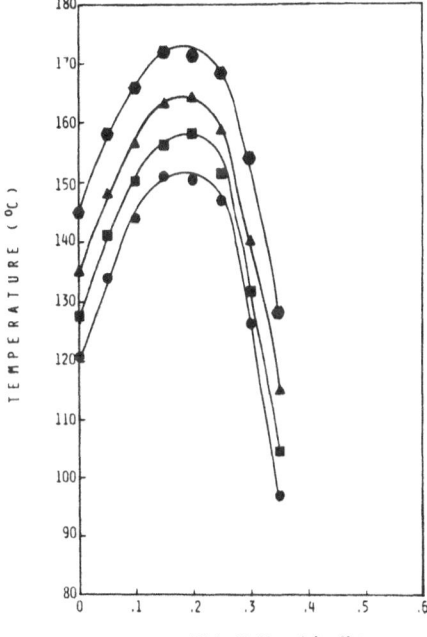

Fig. 6. Miscibility doors of PVME/SMMA blends having different blend ratios (●) 80/20, (■) 50/50, (▲) 35/65, (●) 20/80 [36]

exhibit LCST behavior that also occurs in copolymer based blends. As a result, the miscibility-immiscibility boundary is steeper in the former than in the latter case. Blends of poly(1,4-*cis*-butadiene) (PB) and poly(styrene-*ran*-butadiene) (SB) exhibit in a certain range of copolymer composition both UCST and LCST. The resulting miscibility door is shown in Fig. 7. Similar results have been reported for blends of PS and carbonylated PPO [12].

Table 2. Miscibility doors of PVME with different styrene copolymers

Copolymer	Miscibility domain	Ref.
SAN	AN content $\leq$ 12 wt%, PVME content in the blend 80 wt%, LCST behavior for all miscible blends	[37]
poly(S-*ran*-maleic anhydride) (SMA)	MA content $\leq$ 16 wt%, PVME content in the blend 80 wt%, LCST behavior for all miscible blends	[37]
poly(S-*ran-m*-nitrostyrene)	*m*-nitrostyrene content $\leq$ 30 mol%, LCST behavior for all miscible blends	[38]

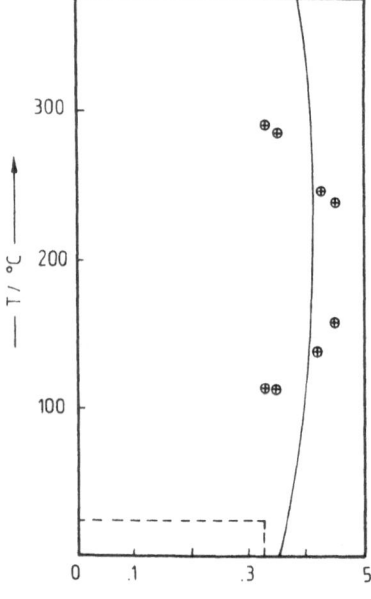

Fig. 7. Miscibility door for 50/50 blends of PB and SB as a function of copolymer composition. The *circles* refer to experimentally determined LCSTs and UCSTs. The *curve* was calculated using the equation-of-state theory discussed in Sect. 2.1. Miscibility occurs to the left of the curve. Inside the *dashed area*, solution cast films are transparent [39]

Miscibility windows originate from a delicate balance of inter- and intramolecular repulsions. Figure 8 gives an example. The corresponding homopolymers are immiscible, but, not so PPO and the random copolymers in a certain range of copolymer composition. Experimentally determined miscibility windows are listed in Table 3.

2) *Blends of random copolymers* $A_x B_{1-x}$ *and* $A_y B_{1-y}$. The phase behavior is significantly determined by the particular microstructure of the constituents, e.g. the sequence distribution [50, 51]. A consistent theory does not exist so far. Experimental results are depicted and summarized in the following overview.

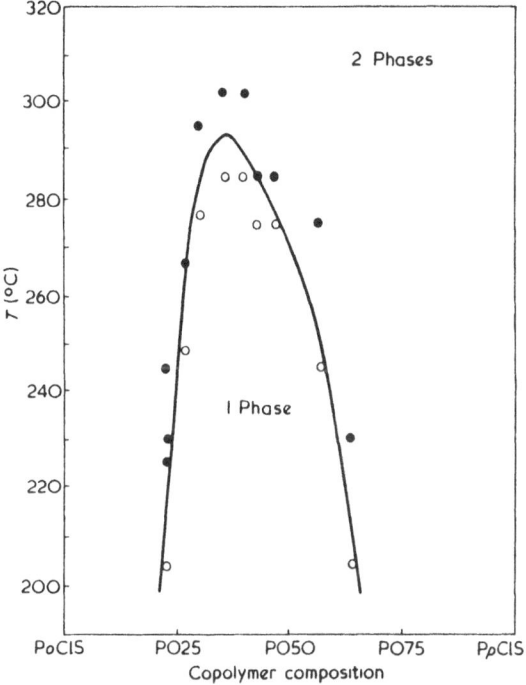

Fig. 8. Miscibility window of 40/60 blends of poly(*o*-chlorostyrene-*ran-p*-chlorostyrene) and PPO; ○ miscible, ● immiscible [40]

I 50/50 blends of $S_x MMA_{1-x}/S_y MMA_{1-y}$, a) at 25 °C, b) at 180 °C; ● one phase, ○ two phases. The solid curves have no theoretical background [52].
    Similar results have been found for 50/50 blends of chlorinated linear polyethylenes [52].

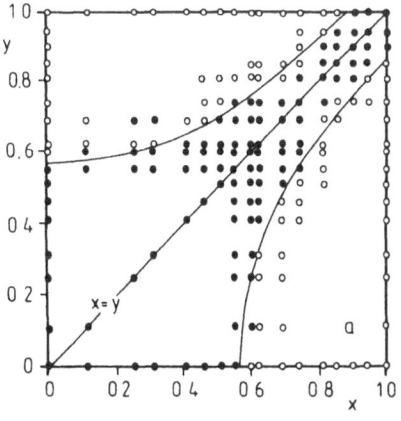
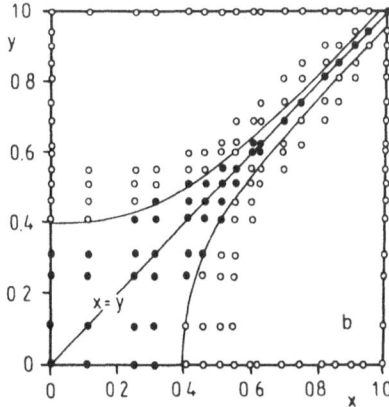

II 50/50 blends of $LC_x MMA_{1-x}/LC_y MMA_{1-y}$ at 25 °C; LC – liquid crystalline unit; ○ two phases, ● isotropic, ▲ nematic, ■ smectic. The solid
curves as in I [52].

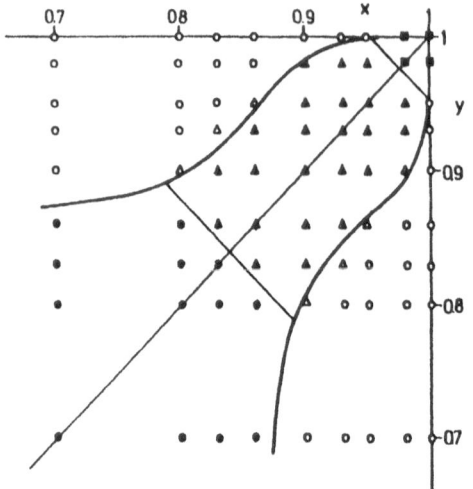

III Copolymers of methyl methacrylate (MMA) and ethyl methacrylate (EMA)
at 70 °C, blending ratio 50/50; ○ one phase, ● two phases. The solid curves
were calculated with $\chi_{AB} = 0.013$ [53]. Similar miscibility maps have been

MMA mol%

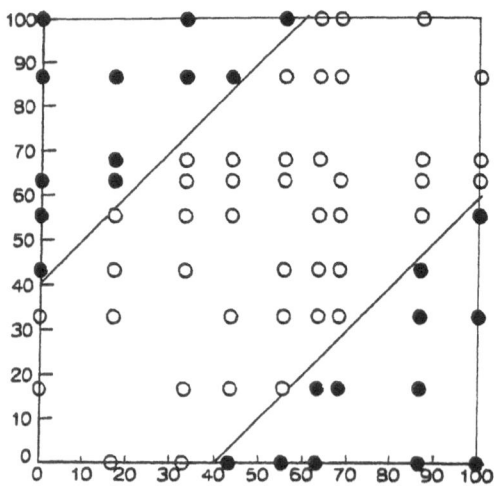

MMA mol%

reported for blends of poly(butyl methacrylate-*ran*-MMA) [52], poly(*N*-phenyl itaconimide-*ran*-MMA) [54], sulfonylated PPO copolymers [55] and even for chlorinated PVCs [53], where the latter result contrasts with findings submitted in [52].

IV 50/50 blends of chlorinated linear polyethylenes at 70 °C; ○ one phase, ● two phases. The dashed curves have no theoretical background [56].
 The result deviates from predictions of the theory in that respect that the miscibility region expands to form bulges. The same phenomenon was observed for 50/50 blends of chlorinated branched polyethylenes (with shrinking miscibility area in comparison to linear polyethylenes) [56] and chlorinated PVCs [52]. Thus, there are conflicting results on chlorinated linear polyethylenes and chlorinated PVCs.

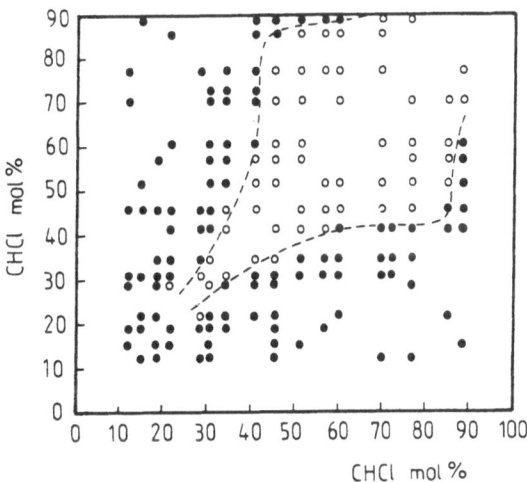

3) *Blends of* $A_x B_{1-x}/A_y C_{1-y}$. The variety of possible miscibility maps enhances tremendously. A systematic classification has been given in Ref. [57]. Typical examples are presented in the following:

 I Random copolymers of vinyl chloride (VC) and vinyl acetate (VAc) blended with chlorinated PVCs at 180 °C, blending ratio 50/50; ○ one phase ● two phases. The solid curve was calculated according to Eq. (26) [57].
 II 50/50 blends of poly(VC-*ran*-VAc) and poly(ethylene-*ran*-VAc) at 160 °C; symbols as in I. The solid and dashed curves were calculated as in I, the latter for infinite molecular masses [57].
III 50/50 blends of SAN and SMMA at room temperature; ● one phase, ○ two phases. The solid and dashed curves were calculated as in II [6].
 Similar miscibility maps were reported for 50/50 blends of SMMA and poly(S-*ran*-maleic anhydride) (SMA) [6]

I

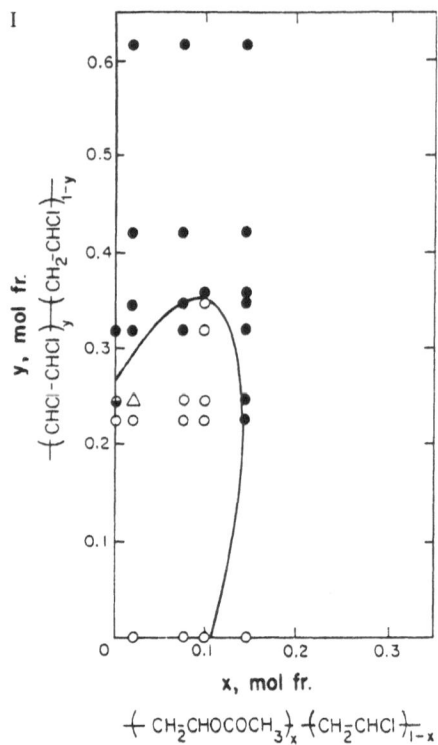

II

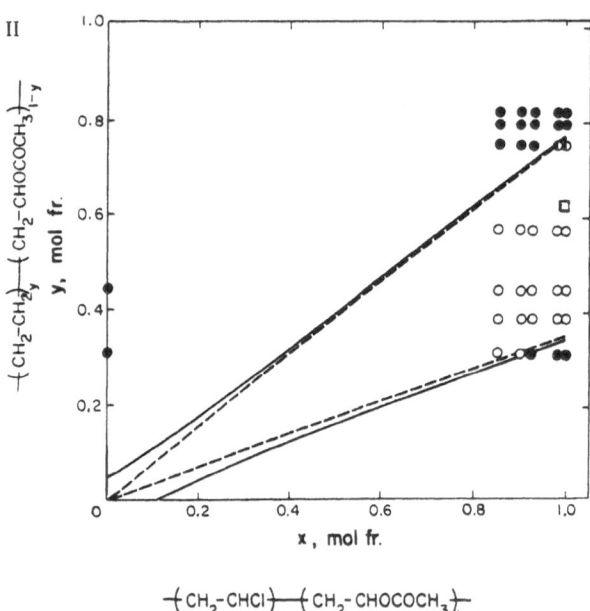

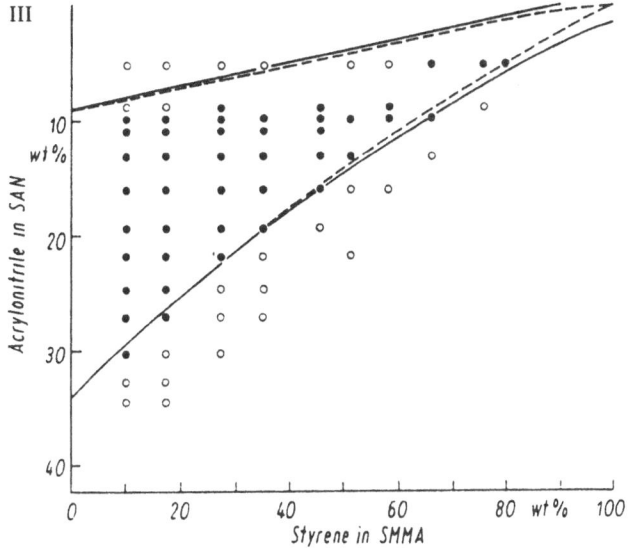

SAN and SMA [58–61]
SAN and poly(S-*ran*-(*N*-phenyl maleimide)) [61]
SAN and poly(*N*-phenyl itaconimide-*ran*-S) (IM-S)
SAN and IM-AN
SMMA and IM-S
AN-MMA and IM-MMA
AN-MMA and IM-AN [54]

IV 50/50 blends of SMMA and MMA-MA; ● two phases. The solid curve is calculated for a degree of polymerization of 500. Both components are miscible in the area above the curve [7].

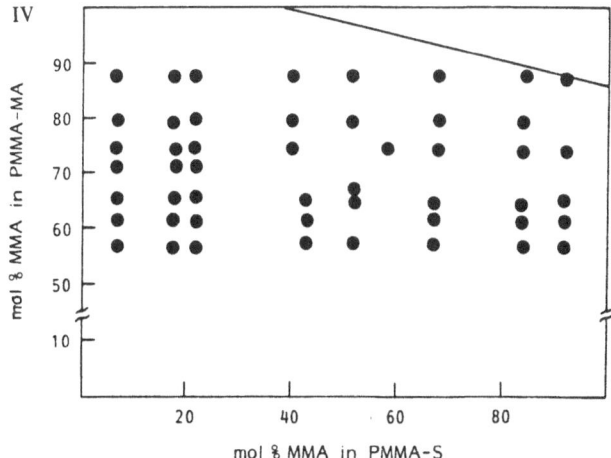

A similar isothermal miscibility map has been reported of the system SMMA/IM-MMA [54].

4) *Blends of two random copolymers without a common segment.* This is the most general case. No systematic classification has been given so far.

I 50/50 blends of sulfonylated PPO (SPPO) with different fluorinated styrene copolymers and SAN, respectively, at 290 °C; experimentally determined miscibility domains inside the solid curves; calculated miscibility regions – shaded areas [55]

  a) SPPO/poly(S-*ran-p*-fluorostyrene)
  b) SPPO/poly(S-*ran-o*-fluorostyrene)
  c) SPPO/SAN

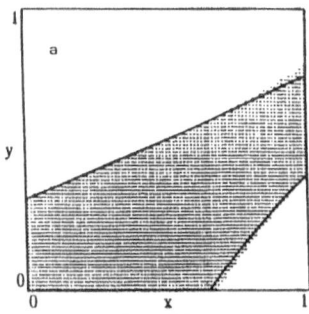
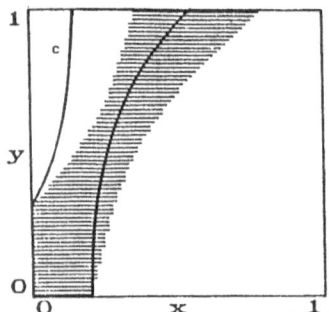

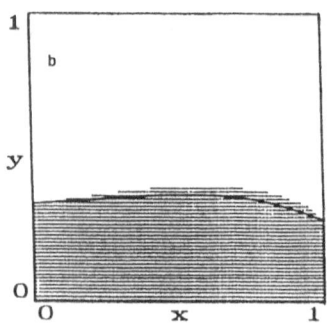

Similar results as in c) have been reported for 50/50 blends of
MMA-AN and SAN [7]
IM-MMA and SAN [62].

II 50/50 blends of MMA-BMA and SAN at 160 °C. The solid curve is calculated for a degree of polymerization of 1000; the dashed lines represent experimental results [63].

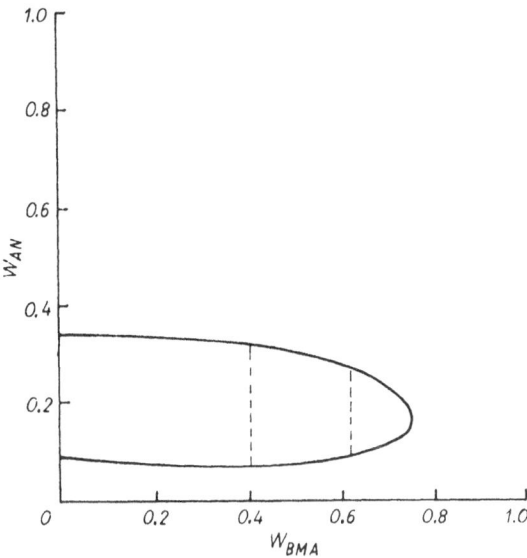

Table 3. Windows of miscibility

Blend	Miscibility domain	Ref.
PPO and poly(o-fluorostyrene-ran-p-chlorostyrene)	miscibility between 5 and 80 mol% of p-chlorostyrene in the copolymer	[29]
PPO and poly (o-chlorostyrene-ran-p-chlorostyrene)	one phase between 20 and 70 mol% p-chlorostyrene	[29]
PPO and poly(o-fluorostyrene-ran-p-fluorostyrene)	one phase region between 7 and 42 mol% p-fluorostyrene	[29]
PMMA/SAN	different authors give the miscibility region between 7–9 wt% and 33–40 wt% of AN in SAN	[5, 41–43]
PMMA/poly(α-methyl styrene-ran-AN)	the border at low α-methyl styrene content is uncertain caused by synthesis problems, the border at high α-methyl styrene content is approximately 40 wt% AN for 60/40 blends	[44]
Poly(ethyl methacrylate)/SAN	miscibility between 6 and 27 wt% AN in SAN	[43]
Poly(n-propyl methacrylate)/SAN	miscibility between 3 and 25 wt% AN in SAN	[43]
Poly(chloro methyl methacrylate)/poly(p-methyl styrene-ran-AN)	miscibility between 13 and 42 wt% AN in the copolymer	[45]
PPO/poly(o-fluorostyrene-ran-p-bromostyrene)	miscibility between 11 and 73 mol% p-bromostyrene	[46]
Poly(ε-caprolactone)/SAN	miscibility between 8 and 28 wt% AN in SAN	[47]
PPO/poly(α-methyl styrene-ran-S)	complete miscibility	[48]
Poly(vinyl chloride)/poly(MMA-ran-butyl methacrylate)	for 50/50 blends complete miscibility up to 190 °C, immiscibility above 240 °C, limited miscibility in between these temperatures	[49]

3 Kinetics of Phase Separation and Phase Dissolution in Polymer Mixtures

3.1 Spinodal Decomposition in Liquid Mixtures of Polymers

In this section we would like to deal with the kinetics of the liquid–liquid phase separation in polymer mixtures and the reverse phenomenon, the isothermal phase dissolution. Let us consider a blend which exhibits LCST behavior and which is initially in the one-phase region. If the temperature is raised setting the initially homogeneous system into the two-phase region then concentration fluctuations become unstable and phase separation starts. The driving force for this process is provided by the gradient of the chemical potential. The kinetics of phase dissolution, on the other hand, can be studied when phase-separated structures are transferred into the one-phase region below the LCST.

The initial wavelength of the growing concentration fluctuations, typical for a phase-separation process in polymer blends, is in the order of 10^3 nm. Therefore, light scattering or optical techniques are suitable for recording this process. Moreover, the high viscosity of the polymers slows down the unmixing to the point that even the early stage of the phase separation process can be observed. In the last decade, the phase separation dynamics in polymer blends has been studied extensively. Since so many studies and reviews have been published, we will mention here only very recent reviews [13, 14, 16, 64]. At present, the concept of spinodal decomposition is generally accepted. A hallmark of this topic is Cahn's linear theory for spinodal decomposition which is still the starting point for both experimental analysis and theoretical development.

Theoretical background. For unmixing in unstable regions there is no energy barrier for the growing concentration fluctuations. According to Cahn [65] this process is considered as belonging to the linear regime, i.e. phase separation is caused by a thermodynamic driving force which initiates a diffusional flux against the concentration gradient. In other words, the unmixing process is treated entirely as diffusion-controlled.

The linear theory predicts that the scattering function $S(q, t)$ which is equivalent to the scattering intensity $I(q, t)$ increases exponentially in the course of time t after initiation of the phase separation

$$S(q, t) = S(q, 0) \cdot e^{2R(q)t} \tag{27}$$

where $R(q)$ is the growth rate of concentration fluctuations. The quantity q is the wave number of growing fluctuations and is also equal to the scattering vector which is a scalar for isotropic samples. It is given by

$$q = \frac{4\pi}{\lambda} \sin \frac{\theta}{2} \tag{28}$$

where λ is the wavelength of the radiation which is employed to study the concentration fluctuations and θ is the scattering angle at which the particular Fourier-component contributes to the scattering intensity. The growth rate $R(q)$ is then

$$R(q) = -Mq^2\left(\frac{\partial^2 G}{\partial \phi^2} + 2Kq^2\right) \qquad (29)$$

where G is the Gibbs free energy of the mixture in which the concentration of one component is given by a constant value ϕ, M is the mobility parameter (i.e. an Onsager coefficient) which is positive and K is the coefficient of the gradient energy term. The thermodynamic driving force is given by Eq. (5). The coefficient K has been also specified in the context of the Flory–Huggins theory [66–68], but the explicit expression will be omitted here.

From Eq. (29) one can easily extract a critical wave number q_c

$$q_c = \left[-\frac{\partial^2 G/\partial \phi^2}{2K}\right]^{1/2} \qquad (30)$$

In the unstable region ($G'' < 0$) for wave numbers $q > q_c$, the growth rate $R(q)$ becomes negative and short wavelength fluctuations [$q = (2\pi/\Lambda)$; Λ being the wavelength of the fluctuation] are damped out whereas long wavelength fluctuations ($q < q_c$) will grow exponentially in time. The growth rate $R(q)$ changes its sign at q_c and reaches a maximum at q_m:

$$q_m = \frac{1}{\sqrt{2}} q_c \qquad (31)$$

Therefore, fluctuations with wavelength $2\pi/q_m$ will grow the fastest. When the phase separation process starts, this characteristic spacing $2\pi/q_m$ appears throughout the system at once. As a result, one observes very regular, highly interconnected two-phase morphologies [20, 69]. Qualitatively, this unique regularity arises from two counteracting processes. Phase separation is driven by diffusion, but, it is simultaneously connected with the establishment of interfaces which enhances the free energy. The tendencies to reduce the interfacial area and to progress the diffusion-controlled phase separation lead to an optimum domain size. The apparent diffusion coefficient D_{app} which governs, according to Cahn, the phase separation process and may be defined by

$$D_{app} = M \cdot \frac{\partial^2 G}{\partial \phi^2} \qquad (32)$$

and is negative for the phase separation process whereas it is positive for the process of phase dissolution. The apparent diffusion coefficient consists of two quantities ruling the phase separation: the kinetic mobility coefficient M and the thermodynamic driving force.

Early stage. Now, the question may arise: Does a period of time exist wherein the growth of concentration fluctuations can be adequately described by Cahn's linearized theory? This approximation is certainly valid only in the limit of small changes in composition which occur in the early stage of the phase separation process and in the small q regime. Let R_0 be the size of a polymer coil and q the wavenumber for a particular Fourier component of growing fluctuations, then the small q domain is characterized by $qR_0 \ll 1$ which means the wavelength of concentration fluctuations is much larger then the coil diameter. The period of time in which the early stage of spinodal decomposition is expected to exist can be approximated by $\tau \sim R_0^2/|D_{app}|$. Thus, the quantity τ describes the time required by a chain molecule to diffuse over a distance comparable to its own size R_0. The large length R_0 and the small diffusion coefficient D_{app} (or the high viscosity) of polymeric systems make the characteristic time τ very large in comparison to small-molecule systems (order of magnitude: 10^3 s). Obviously, that is the reason that experimental evidence exists for the approximate validity of the linearized theory for polymer mixtures in their early stage of phase decomposition and in the small q limit. Cahn's theory as described in Eqs. (27) and (29) throws three bridges to experimental examination:

– the time evolution of scattering intensity is exponential at q = const according to Eq. (27);
– the intensity I as a function of q at t = const should display a maximum at q_m; the position of the maximum should be independent of time: $q_m(t) = q_m(0) \cdot t^0$ (cf. Fig. 9);
– in the $R/q^2 - q^2$ plane Eq. (29) should become a linear relationship

$$\frac{R(q)}{q^2} = - D_{app}\left(1 - \frac{q^2}{q_c^2}\right) \tag{33}$$

where $\lim_{q \to 0} \dfrac{R(q)}{q^2} = - D_{app}$.

As a particular stringent test of the linearized theory, we must considered the linear variation of R/q^2 with q^2 and failure of this linearity has been suggested as arising from shortcomings in the theory [70–73].

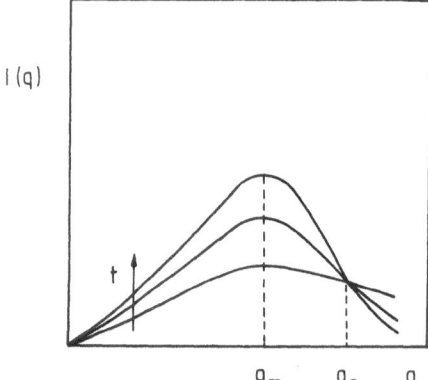

Fig. 9. Schematic representation of scattering intensity as a function of wave number q in the early stage of phase separation where $q_m \sim t^0$. This agrees with the linear theory

The characteristic parameters D_{app} and q_c occurring in the theory can be expressed in the context of the mean-field model for polymer mixtures by [74, 75]

$$D_{app} = D\varepsilon, \quad q_c^2 = 2q_m^2(0) = \frac{3}{R_g^2}\varepsilon \quad \text{with} \quad \varepsilon \equiv \frac{\chi}{\chi_s} - 1 \sim \Delta T \qquad (34)$$

D – self-diffusion, R_g – radius of gyration.

The quantity χ_s is the thermodynamic interaction parameter χ between the constituent polymers at the stability limit and $\Delta T \equiv (T - T_s)$ is the quench depth.

Most of the observations reported so far suggest, that there are certain ranges of time, quench depth and wave numbers q where unmixing progresses to a good approximation in the linear branch, i.e. non-linearity is very weak. The conditions for validity of the linear theory can be specified more precisely as follows:

1) small q domain, i.e. $q \ll q_c$;
2) sufficiently large quench depth ΔT or ε, i.e. $|\varepsilon| \gg (1/r)$;
3) the period of time is confined to the range where thermal noise does not significantly affect the growth of fluctuations and the onset of coarsening processes.

Thermal fluctuations can contribute dominantly to the scattering intensity right after the isothermal phase separation starts [70, 76]. Therefore, conditions 1) and 3) must be fulfilled to ensure that the effect of thermal noise is negligible. The dynamics of phase separation can be adequately described by the mean-field model if condition 2) is satisfied. Condition 2) is a direct consequence of the Landau–Ginzburg criterion [75]. Thus, one may establish: prerequisites for Eqs. (27) and (33) are the conditions 1) and 3), while Eq. (34) requires conditions 2) and 3). For example, Eq. (27) and as a consequence Eq. (33) cannot be confirmed experimentally not even for small values of q if the quench depth ε is too small [70]. Moreover, owing to the effect of thermal fluctuations, Eq. (33) fails at $q \approx q_c$ even if the Landau–Ginzburg criterion is fulfilled [70, 77]. Thus, in the former case condition 2) is violated whereas in the latter example conditions 1) and 3) are not satisfied.

The validity of the linear theory observed for the early stage of spinodal decomposition is chiefly related to the large size of the chain molecules. As shown above, characteristic quantities as the time τ or the wavelength $\Lambda_m(0)$ of the fastest growing fluctuation are proportional to R_0^2 and R_g, respectively. Furthermore, the Landau–Ginzburg criterion (cf. condition 2)) ensures that the mean-field regime is sufficiently extended.

Late stage. The consequence of Cahn's linearized theory is that the growth of dominant concentration fluctuations is ruled by $q_m \sim t^0$, i.e. it takes place with no change in the size. Experimental results show that this is fulfilled in the early stage of phase separation. In later stages when non-linear effects become operative both amplitude and size of the concentration fluctuations grow as it is shown schematically in Fig. 10. The evolution of phase structures involves the growth of domain size whereby structures change self-similarly. Therefore, a time-dependent heterogeneity length $L_m(t) \sim \dfrac{1}{q_m(t)}$ of the phase structure exists which is expected to follow a power law

$$L_m(t) \sim t^\alpha \qquad (35)$$

while for the scaling behavior of the peak intensity it has been proposed that

$$I_m(t) \sim t^\beta. \qquad (36)$$

Self-similarity of the phase structure in the late stage leads to the scaling law for the structure function

$$S(q, t) = q_m^{-d} \cdot \bar{S}(q/q_m) \qquad (37)$$

where $\bar{S}(x)$ is a universal function being independent of time t and d is the spatial dimensionality. Comparing Eqs. (35)–(37) it follows that for $d = 3$

$$\beta = 3\alpha$$

The following expressions for $\bar{S}(x)$ have been proposed for an unstable critical and off-critical mixture, respectively [83]

$$\bar{S}(x) \sim \frac{x^2}{3 + x^8} \text{ (critical)} \qquad \bar{S} \sim \frac{x^2}{2 + x^6} \text{ (off-critical)} \qquad (38)$$

Note that for $x \gg 1$ it follows that

$$\bar{S}(x) \sim x^{-6} \text{ (critical)} \qquad \bar{S}(x) \sim x^{-4} \text{ (off-critical)} \qquad (39)$$

The exponent α indicates the growth mechanism which is ruled chiefly by

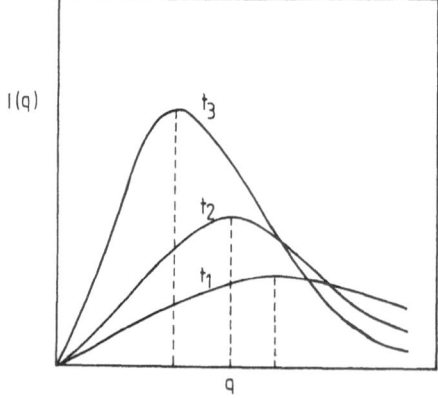

Fig. 10. Scattering intensity as in Fig. 9, but, for the late stage, i.e. in the non-linear branch where q_m shifts to smaller values in the course of time $(t_1 < t_2 < t_3)$

diffusion, viscosity, interfacial tension and curvature and volume ratio of unmixed phases.

Various theoretical attempts have been made to describe the later stages of spinodal decomposition. The validity of the scaling laws given by Eqs. (35) and (36) results as a common feature of all these models. However, they differ in the prediction of the exponents α and β. Time-independent values, $\alpha = 0.212$ and $\beta = 0.81$ or $\alpha = 1/3$ and $\beta = 1$, were submitted in Refs. [78] and [79], respectively. More recently, it has been proposed that the scaling exponents themselves should be time-dependent [80–82].

Two growth mechanisms, becoming operative after the early stage, will be mentioned here. When the growth mechanism in later stages is governed by hydrodynamic interactions and the Stokes formula $D \sim kT/L\eta$ for the diffusion coefficient may be applied then it follows that

$$L^2 \sim Dt \sim \frac{kT}{L\eta} \cdot t$$

and

$$L \sim \left(\frac{kT}{\eta}\right)^{1/3} t^{1/3}: \quad \alpha = 1/3 \tag{40}$$

which is equivalent to the Lifschitz–Slyozov process [84].

The long-term coarsening behavior in the late stage is driven by the interfacial tension σ at a rate controlled by the viscosity η [85, 69]. It follows that

$$L \sim \frac{\sigma}{\eta} \cdot t: \quad \alpha = 1 \tag{41}$$

From all theoretical and experimental results one may conclude that there is no simple scaling relation over a long period of time. The detailed coarsening mechanisms, which are attributed to the intrinsical non-linearity of the phase separation process, determine the exponent α. The time dependence of α reflects cross-over among different coarsening processes.

The structure function $\bar{S}(q/q_m)$ given in Eqs. (38) and (39) turns out to be proportional to $(q/q_m)^{-4}$ at $q > q_m$ for both critical and off-critical mixtures [64, 86]. This is consistent with the theoretical result for off-critical mixtures (cf. Eq. (39)) but not for critical mixtures.

Two effects attributed to long-chain molecules cause features of the coarsening processes taking place in phase-separating polymer systems:

– molecular entanglements
– the thickness of the interfacial region between two coexisting phases.

Entanglements severely influence the viscosity which in turn affects significantly the dynamics of the unmixing process. The interfacial layer of polymer–polymer systems is more extended than for small-molecule systems, but, is smaller than the size of a chain molecule. Consequently, the interfacial tension in polymer systems is significantly influenced by the loss of conformational entropy of chain molecules in the interfacial region. An effect again typical for polymer systems that does not exist in small-molecule systems.

Phase dissolution in polymer blends. The reverse process of phase separation is phase dissolution. Without loss of general validity, one may assume again that blends display LCST behavior. The primary objective is to study the kinetics of isothermal phase dissolution of phase-separated structures after a rapid temperature-jump from the two-phase region into the one-phase region below the lower critical solution temperature. Hence, phase-separated structures are dissolved by a continuous descent of the thermodynamic driving force responsible for the phase separation. The theory of phase separation may also be used to discuss the dynamics of phase dissolution. However, unlike the case of phase separation, the linearized theory now describes the late stage of phase dissolution where concentration gradients are sufficiently small. In the context of the Cahn theory, it follows for the decay rate $R(q)$ of Eq. (29) [74]

$$R(q) = Mq^2 \cdot \left[\frac{\chi - \chi_s}{\chi_s} + \frac{R_0^2 q^2}{36} \right] \tag{42}$$

For an experimental situation where R_0 and q are in the order of $100 \, \text{Å}$ and $10^5 \, \text{cm}^{-1}$, respectively, it follows that $R_0^2 q^2/36 = 0(10^{-4})$ (0 being the symbol for order of magnitude) and also the criterion for the small-q regime is satisfied. Moreover, $(\chi - \chi_s)/\chi_s \approx (T - T_s)/T_s > 0(10^{-2})$ results for all blends studied below. Under the small-q regime covered here one arrives at

$$\frac{\chi - \chi_s}{\chi_s} \gg \frac{R_0^2 q^2}{36}. \tag{43}$$

Consequently, the gradient free-energy term may be neglected. After the temperature jump from the two-phase region into the one-phase region below LCST the phase dissolution starts leading finally to a homogeneous system. Therefore, below LCST the scattering intensity I decreases as the phase dissolution progresses with the time of annealing. The intensity decay reflects the phase dissolution. According to Eq. (27) the intensity descends exponentially with time

$$\ln \frac{I}{I_0} = -2D_{app} q^2 t \tag{44}$$

where t and I_0 are the annealing time after the temperature jump and the scattered intensity at annealing time zero, respectively. In Eq. (44) the relations (32) and (43) have been used. This simplified model allows one to characterize the dynamics of phase dissolution in terms of the apparent diffusity D_{app}.

In the case of phase-separation dynamics, the non-linear terms become increasingly important in the course of time. In contrast, for phase-dissolution kinetics, the non-linear terms are most important in the beginning and their importance declines with progressing dissolution.

One has to add here that for determination of D_{app} from Eq. (44) it is convenient to replace the quotient I/I_0 by the corrected expression $(I - I_\infty)/(I_0 - I_\infty)$ where I_∞ is the scattered intensity in the stable state below LCST at the corresponding temperature to facilitate the analysis based on Eq. (44).

From Eq. (32) it is obvious that the apparent diffusion coefficient comprises a kinetic aspect, M, as well as a thermodynamic aspect, $\partial^2 G/\partial\phi^2$. If one recognizes that the dissolution process starts after a rapid temperature jump has taken place from the two-phase into the one-phase region then one may say to a good approximation that M is related to the final state of the temperature jump while $\partial^2 G/\partial\phi^2$ characterizes the initial state before the temperature jump. Therefore, if one chooses different experimental courses as shown in Fig. 11 one can approximately reveal the different aspects of D_{app}. Temperature jumps from one single temperature in the two-phase region to different temperatures in the homogeneous region below LCST (Fig. 11a) correspond to a variation of the mobility M as a function of temperature whereby the thermodynamic driving force is kept approximately constant. The course indicated in Fig. 11b, on the other hand, focuses attention on the influence of the thermodynamic driving force at constant mobility.

Results are shown in Figs. 12 and 13. All blend specimens were set isothermally above LCST and kept there for a maximum of 5 min. As will be seen, this corresponds only in some cases to an early stage of spinodal decomposition depending on temperature. The diffusion coefficients governing the dynamics of phase dissolution below LCST are in the order of 10^{-14} cm^2 s^{-1}. Figure 12 reflects the influence of the mobility coefficient on the phase dissolution. As can be seen, the apparent diffusion coefficient increases with increasing temperature of phase dissolution which expresses primarily the temperature dependence of the mobility coefficient. Furthermore, it becomes evident that the mobility obeys an Arrhenius-type equation. Similar results have been reported for phase dis-

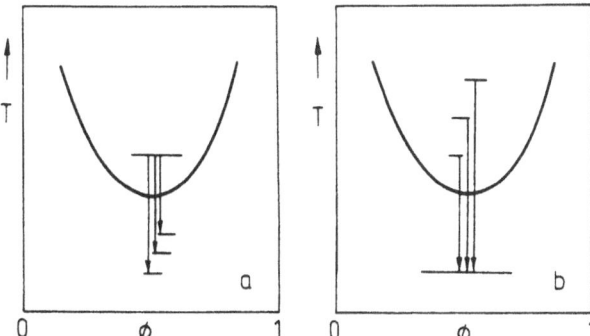

Fig. 11a, b. Courses of the experimental procedure in phase dissolution experiments. **a** The thermodynamic driving force is kept constant. **b** The mobility remains constant

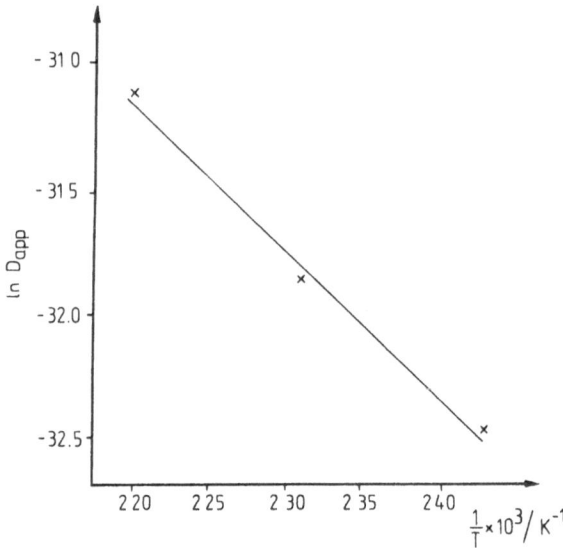

Fig. 12. Arrhenius plot of the apparent diffusion coefficient for PMMA/SAN-31.5 (50/50)(31.5 wt%
AN in SAN). The apparent diffusion coefficients results after temperature jumps from 210 °C to
different annealing temperatures below the LCST (cf. Fig. 11a). Phase separation of the blend starts
at 200 °C

solution in blends of *cis*-1,4-polybutadiene and poly(styrene-*co*-butadiene) [88].
Also for blends of polystyrene and poly(vinyl methyl ether) the apparent diffusity for phase dissolution descends with decreasing temperature, but, conflicting
results have been submitted on the temperature dependence. Arrhenius-type
behavior was found in [89] while it could not be confirmed in [90]. Contrary to
these results, for polyisoprene blended with poly(vinyl ethylene-*co*-1,4-
butadiene) the apparent diffusity of phase dissolution has been found to increase
with decreasing temperature according to an Arrhenius relation [91]. This
behavior is not obvious at present.

   A completely different result comes out if one pursues the experimental
course of Fig. 11b. The variation of the apparent diffusion coefficient as a function of quench depth $\Delta T \equiv T - T_s$, where T_s is the spinodal point, is shown in
Fig. 13. Initially, the apparent diffusity increases with increasing quench depth
ΔT which is in accord with the mean-field theory since $(\chi - \chi_s)/\chi_s \sim \Delta T$.
However, when the quench depth is raised further the apparent diffusion
coefficient starts to decrease and finally, apparently levels off.

   This behavior may be caused by the fact that for a large enough ΔT phase
separation proceeds beyond the early stage, generating phase-separated morphologies typical for a late stage of spinodal decomposition. Consequently, the
reverse process of phase dissolution is not entirely diffusion-controlled. Although one still observes, to a good approximation, an exponential decay of the
scattered intensity over an extended period of time, Eq. (44) has just a formal

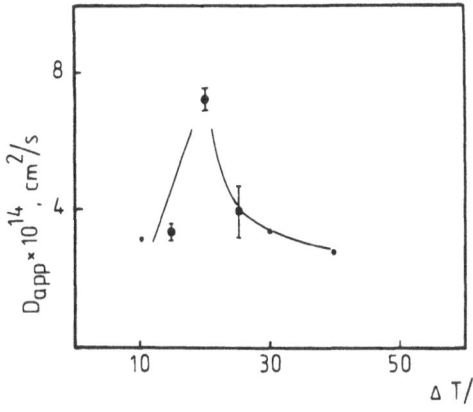

Fig. 13. Apparent diffusity as a function of quench depth for the system as in Fig. 12 and the procedure as indicated in Fig. 11b. Annealing temperature below LCST: 190 °C

meaning, however. For a certain period of time the thermodynamic driving force of phase dissolution is also ruled by the interfacial tension. Thus, the latter effect and diffusion are mingled in the "diffusity" D_{app} of Eq. (44) which causes the behavior shown in Fig. 13.

3.2 Ordering Dynamics and Morphology Control

The unmixing mechanisms and their kinetics, discussed before, offer the possibility of morphology control at different spatial levels. Various homogeneous blends transferred into the thermodynamically unstable region display, as a result of the phase separation process via spinodal decomposition, regular, network-like two-phase morphologies. Similar phenomena can be observed in solutions of immiscible polymers when phase separation is induced by solvent evaporation or coagulation. Depending on the interactions between the polymers and the rate of solvent evaporation, the time evolution of concentration fluctuations leads to bicontinuous, network-like, phase-separated structures or to two-phase morphologies having irregular shape and size of domains. These phenomena naturally provoke us into developing principles of morphology control in polymer blend systems through phase changes. Especially challenging is to be able to analyse kinetically regulated stability limits for ordered microdomain morphologies.

In polymer blends transferred into the unstable region, the unmixing process is chiefly governed by temperature and the corresponding interactions between the components. Periodic concentration fluctuations develop in the early stage and are diffusion-controlled. These demixing structures may turn in later stages into regularly ordered network-like morphologies. Phase evolution in the late stages is driven by the interfacial tension at rates controlled mainly by the viscosity of the phases. The phases then have compositions corresponding to the miscibility gap. As various observations revealed, the periodic structures can grow self-similarly over a certain period of time. Afterwards coalescence and break-up of entire phase domains take place resulting in irregularly shaped morphologies. Similar phenomena can be also observed in blends of highly immiscible polymers seized in forced mixtures. Above a certain temperature when at least one of the components becomes mobile phase separation occurs.

Ternary solutions of immiscible polymers in a low-molecular solvent display wide miscibility gaps. Consequently, they invariably involve demixing above a critical concentration of total polymer by spinodal decomposition and subsequent coarsening processes. When solvent evaporation progresses the enhanced viscosity will slow down the rate of phase separation to a level at which no further phase changes can be observed.

The following discussion will be restricted to evolution of phase morphologics preferably in late stages in solutions and blends of immiscible polymers wherein phase separation is initiated by solvent evaporation during casting and thermal agitation, respectively.

Blend solutions. Solutions of blends comprising immiscible polymers P_1 and P_2 in a nonselective solvent have miscibility gaps as shown schematically in Fig. 14. When the polymer concentration increases by solvent evaporation the polymer coils start to interpenetrate above a certain concentration. As a consequence, interactions between the polymers become operative and phase separation must start above a critical polymer concentration ϕ_p^c. The composition of the new phases will be situated on the branches of the coexistence curve. Finally, the unmixing process is arrested owing to enhanced viscosity. This simple scheme reveals the factors directing morphology evolution in blend solutions:

– the rate of solvent evaporation which determines the total polymer concentration ϕ_p where phase separation starts;

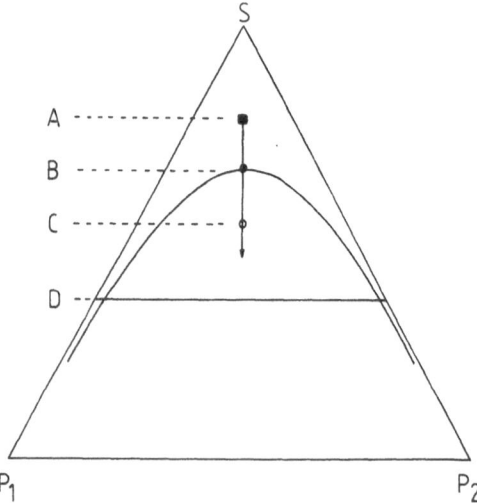

Fig. 14. Schematic phase diagram of a blend solution of immiscible polymers P_1 and P_2. A – initial blend solution, B – stability limit, C – onset of phase separation, D – slowing down of phase separation owing to enhanced viscosity

- the polymer–polymer interaction which influences the critical polymer concentration ϕ_p^c;
- the blend ratio of polymers.

It is obvious that the difference $(\phi_p - \phi_p^c)$ is closely related to the "quench depth" or the thermodynamic driving force while the rate of the process is determined by the viscosity. According to these factors structure evolution is ruled by both externally imposed conditions and intrinsic properties of the system.

Morphology evolution during solution casting of polymer blends has been studied to some extent in recent years [93–96]. More than 10 blend solutions comprising immiscible polymers and random copolymers such as poly(methyl methacrylate) and different styren-based copolymers were studied in Ref. [93]. The results can be summarized as follows:

- the faster the solvent evaporation, the smaller the periodic distance of the network-like morphology;
- the periodic distance of the regular morphology descends with increasing deviation of the blend ratio from symmetry (50/50 blend);
- some blend solutions as PMMA/PS dissolved in toluene failed to form periodic morphologies even at high evaporation rates.

Similar results have been submitted on solutions of PMMA and poly(styrene-co-acrylonitrile) (SAN) in toluene [96]. The components are miscible when the AN content in SAN ranges from 9.4 up to 34.4 wt% (window of miscibility). Only immiscible pairs were studied. An example of structure evolution in blend solutions, comprising copolymers of different AN content, in the course of annealing time at constant annealing temperature is shown in Fig. 15. The results can be summarized as follows:

- regular morphologies develop only in the immediate vicinity of the window of miscibility;
- these morphologies show increasing periodic spacing with both annealing time and temperature. The latter effect seems to be contrary to the result submitted in Ref. [93].
- deviations from symmetry result in smaller periodic spacings;
- in a sufficient distance from the window of miscibility solution casting of the incompatible polymers leads to two-phase morphologies having irregular shape and size. Moreover, unmixing elapses very rapidly. It has essentially ceased after 1 min.

Morphology evolution has been also studied in blend solutions of cellulose and aramide dissolved in dimethyl acetamide/LiCl as solvent [94, 95]. Selected examples are shown in Fig. 16. The initially transparent samples were kept for certain periods τ at room temperature and afterwards annealed at temperatures ranging from 60 up to 130 °C until no further phase changes could be detected.

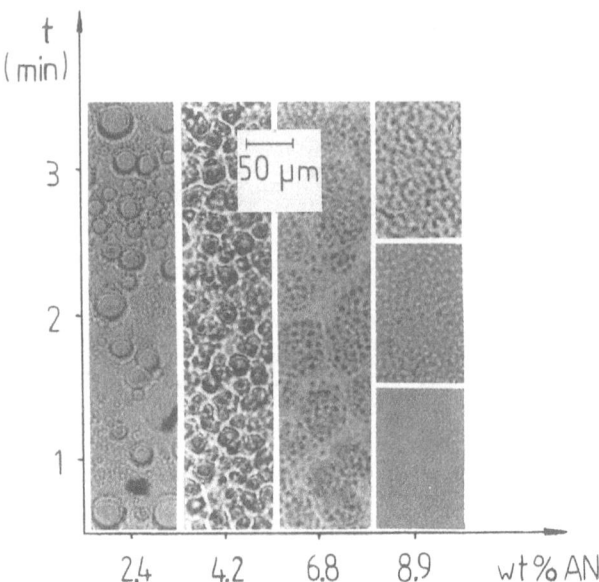

Fig. 15. Optical micrographs of phase decomposition morphologies obtained from PMMA/SAN blend solutions of different AN content in the course of annealing time; annealing temperature: 70 °C

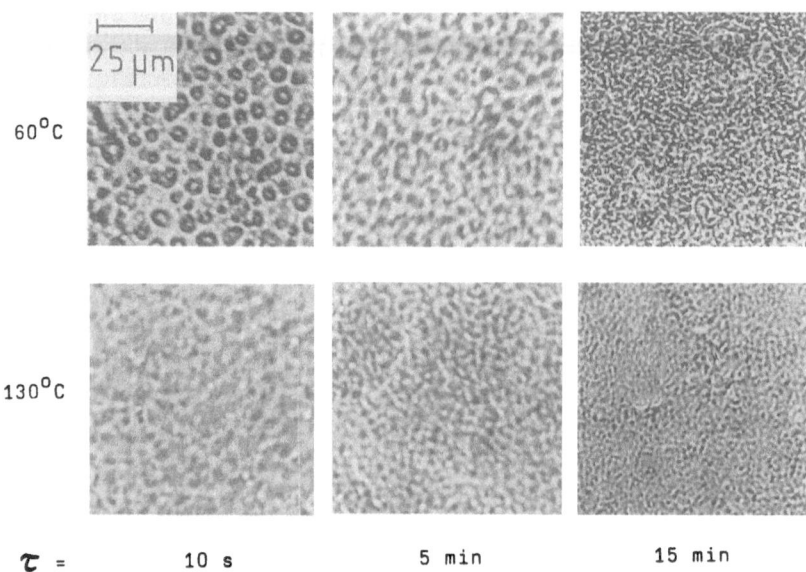

Fig. 16. Optical micrographs of decomposition structures obtained from 50/50 blend solutions containing initially 3 wt% of total polymer. Storage periods τ at room temperature and annealing temperature as indicated

The results can be summarized qualitatively as follows:

– regular morphologies develop at low and high evaporation rates;
– moderate evaporation rates lead to irregularly formed, coarse-grained phase structures;
– solutions having aramide in excess display finer periodic phase-structures than solutions with cellulose in excess;
– for solutions kept for 10 s at room temperature and annealed afterwards at 60 °C the unmixing process ceases after about 5 min. This process progresses much more rapidly and can be finished within 2 min for more extended storage periods coupled with higher annealing temperatures.
– the differences in the morphologies resulting from different evaporation courses are less pronounced for solutions having initial polymer concentrations near to the critical concentration. More diluted solutions tend to more coarse-grained structures than the concentrated solutions.

Polymer blends. Morphology control has been also studied in binary polymer blends comprising highly immiscible polymers. Forced mixtures of isotactic polypropylene (PP) and ethylene-propylene random copolymer (EPR) prepared by rapid coagulation from solutions were discussed in Refs. [19, 97]. The thermal history for a morphology control involves demixing of the mixture above T_m of PP in the mixture for a certain period of time and subsequent crystallization of the melts subjected to unmixing by quenching below T_m. Thus, there exists an interplay of two kinds of phase transactions: The liquid–liquid (L–L) phase separation above the melting temperature T_m and the liquid–solid transition below T_m, at which the rate of crystallization is much faster than the rate of L–L unmixing. L–L unmixing results in a periodic spacing d of the regular phase pattern and the subsequent crystallization process leads to an average size D of the spherulites. Thus, one may distinguish two cases: $d < D$ and $d > D$. When isothermal unmixing in the melt proceeds for a certain period of time, crystallization occurs in the PP-rich domains without involving segregation of PP and EPR during crystallization. In this way the L–L demixing patterns can be stored within the spherulites ($d < D$).

When the L–L unmixing process lasts for a long enough period of time, coarsening of the regular morphology advances and d becomes so large that several nuclei arise within a PP-rich domain when the temperature is set below the melting temperature. Thus, the PP-rich regions are transferred into "spherulitic" domains. These domains and the almost amorphous EPR-rich domains form the regular pattern preserved from the L–L unmixing process, but now $d > D$ holds. The superposition of these phase changes provides an excellent opportunity for maintaining various L–L demixing memories during rapid crystallization. Combined effects of crystallization and phase separation on morphology formation has recently been also investigated in blends of poly(ε-caprolactone) and oligomeric polystyrene [98].

Morphology evolution in binary mixtures containing liquid-crystalline polymers have been the subject of several studies in recent years [21, 99, 100]. The system used was a 50/50 mixture of commercial poly(ethylene terephthalate) (PET) blended with the copolyester comprising 60 mol% p-oxybenzoate units and 40 mol% ethylene terephthalate units (P(OB-co-ET)-60). The copolyester exhibits thermotropic liquid-crystalline behavior. In the blend three phase transitions compete:

- the solid–liquid transition
- the liquid–liquid phase separation
- the formation of an anisotropic phase.

When the blend is heated up melting of the PET-crystallites turns out to be the key process. This process is simultaneously the onset process for liquid–liquid phase separation which elapses rapidly. The liquid–liquid phase separation in turn permits the formation of the anisotropic phase. When the PET content of the blends exceeds 60% the formation of the anisotropic phase is hampered and finally impossible.

Special features of the isothermal unmixing process are the following:

1) in the course of phase separation periodic two-phase structures emanate; their characteristic spacing increases as the phase separation advances;
2) owing to the incompatibility of the two constituents the system decays in an anisotropic phase, which is rich in liquid-crystalline polymer and an isotropic phase formed chiefly by the isotropic liquid polymer;
3) structures grow self-similarly over a certain period of time afterwards coalescence occurs.

Conflicting results have been found for the explicit time evolution of the correlation length during isothermal phase separation. A 1/3-power law in the growth of patterns, which is characteristic for the hydrodynamically controlled Lifshitz–Slyozov process, was confirmed in Ref. [99] while an exponential increase over a certain period of time was established in Ref. [21]. Nevertheless, it is evidenced that in blends comprising liquid-crystalline polymers spinodal decomposition and subsequent coarsening processes take a course similarly to isotropic liquid mixtures.

Equivalent results have been reported on blends of poly(p-phenylene benzobisthiazole) (PBZT) and polyamide-6.6 (PA-6.6) [101] and on poly (p-phenylene terepthalamide)(PPTA) blended either with PA-6.6 or polyamide-6 (PA-6) [102, 103]. But, also differences occur in the phase behavior which originate from the fact that these systems are combinations of stiff chain and flexible coil molecules. The high anisotropy of the rigid-rod component may influence the dynamics of the phase separation process. In the system PBZT/PA-6.6 (45/55) the unmixing process initially progresses rapidly at temperatures above the melting point of PA-6.6. Lateron the domain growth slows down dramatically and coalescence cannot take place. Once domains of the rigid-rod component are formed, which must be considered as a solid phase, further rearrangements of the molecules or coalescence of domains cease. This behavior contrasts with blends comprising flexible coil molecules which undergo liquid–liquid phase separation leading finally to coalescence of regular morphologies. Thus, unmixing processes in blends of rigid-rod and flexible coil molecules seized in forced mixtures are not true liquid–liquid phase separations – although similar over a certain period of time – since one component is not fusible.

Morphology evolution as a function of blend composition was studied in forced mixtures containing two of the components PPTA, poly(p-phenylene 1,3,4 oxadiazole) (p-PODZ) and PA-6 [103, 104]. Neither PPTA nor p-PODZ are fusible. Thin, transparent film specimens of the respective polymer blends were obtained by rapid coagulation in distilled water from solutions in sulfuric acid. Optical microscopy of the dried films did not reveal any phase separation. One has to add here, the term "forced mixture" does not necessarily mean a dispersion of the components on a molecular scale has been formed throughout. However, as optical inspection shows, at least a very fine dispersion of the components may be assumed. For a p-PODZ/PPTA 70/30 blend electron microscopy verified this fact. Large scale phase separation can be prevented under the coagulation conditions used. The size of the particles amounted to 100 nm at maximum. Selected examples of PPTA/PA-6 blend morphologies as they evolved after thermally induced phase separation are shown in Fig. 17. These phase patterns develop when the phase separation proceeds isothermally slightly above the corresponding melting temperature for approximately one minute. Afterwards the rate of phase separation slows down and the process does not progress further. Blends having PPTA contents of 70% and more did not show phase separation. At high PPTA content the mobility of PA-6 is severly restricted. The matrix formed by stiff chains "stabilizes" the PA-6 droplets and prevents coagulation. With increasing PA-6 proportion network-like phase structures grow in the course of phase separation which again are stabilized by the nearly immovable but percolating skeleton of the stiff chains. This effect lessens naturally with increasing PA-6 content.

A slightly different behavior can be observed for p-PODZ/PA-6 blends. Owing to the higher mobility of p-PODZ molecules compared to PPTA chains regular structures can be formed at high p-PODZ contents. But these structures rapidly decay into irregularly shaped morphologies with increasing PA-6 content. However, regular patterns develop again at the other end of the composition scale in blends containing 20 or 10 wt% of p-PODZ.

Phase separation induced by thermal agitation cannot be observed in p-PODZ/PPTA blends up to 400 °C owing to the infusibility of both components. Phase morphologies formed in the coagulation process are maintained at higher temperatures. Different phase structures can only be obtained by variation of the coagulation conditions. Regular, network-like phase patterns start to develop when the H_2SO_4 concentration of the coagulation bath exceeds 30%. With increasing H_2SO_4 concentration the periodic spacings increase.

Qualitative analysis of percolation limits. Experimental results suggest that periodicity of the growing morphologies exists only over a certain period of

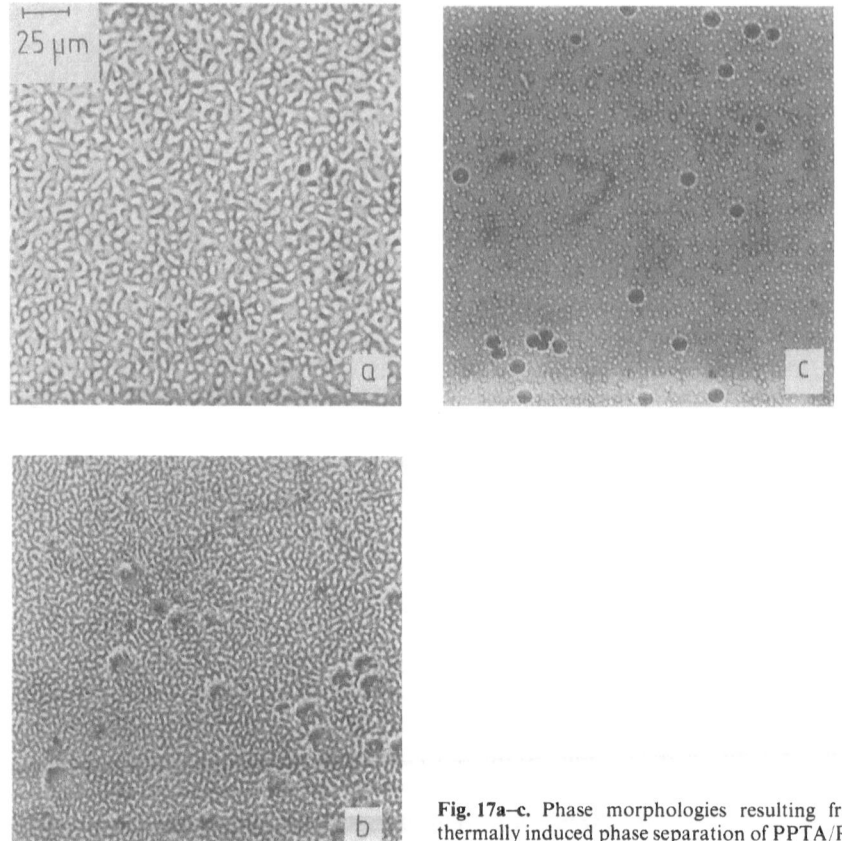

Fig. 17a–c. Phase morphologies resulting from thermally induced phase separation of PPTA/PA-6 blends, **a** 25/75, **b** 40/60 **c** 60/40

time. This leads us to characterize percolation limits by the period of time after which continuous morphologies decay in the course of phase separation into droplet-matrix structures. Hence, percolation limits are equivalent to the existence of critical domain sizes d_c. One may say, the phase-separated structure grows self-similarly over a range of length scales bounded by an upper critical cut-off length which is determined by inner parameters of the system. Above d_c it breaks up into an irregularly shaped morphology.

The simplest expectation is that the critical length d_c is attained when the equilibrium interfacial tension σ is approached in the course of phase decomposition:

$$d_c = \left(\frac{kT}{\sigma}\right)^{1/2} \tag{45}$$

where kT is the thermal energy. According to the mean-field theory the interfacial tension of a phase-separated blend solution of total polymer concentration ϕ_p is given by [105, 106]

$$\sigma \sim kT \phi_p^{1/2} \cdot \varepsilon^{1/2} \qquad \text{with} \quad \varepsilon \equiv \frac{\chi}{\chi_c} - 1 \qquad (46)$$

where χ_c represents the interaction parameter χ of the polymers in the solution at the critical point. Again in the context of the mean-field approximation the quantity ε may be expressed by

$$\varepsilon = \frac{\phi_p}{\phi_p^c} - 1 \qquad (47)$$

Here, ϕ_p^c is the total polymer concentration at the critical limit which reads

$$\phi_p^c = \frac{1}{2r \chi_{AB} \theta_A \theta_B} \qquad (48)$$

r – degree of polymerization
χ_{AB} – interaction parameter of the constituent polymers
θ_i – relative volume fraction of polymer component i
Thus, one arrives at

$$d_c \sim \frac{1}{\phi_p^{1/4} \cdot \varepsilon^{1/4}} \qquad (49)$$

If one assumes that domain growth proceeds below d_c in conformity with the Lifschitz–Slyozov process (Eq. 40), then, one can associate d_c with the period of time t_c in which the periodic structure exists after initiation of the phase separation. This may be slightly generalized to $d \sim t^\alpha$. It follows that

$$t_c \sim \frac{\eta}{kT} \frac{1}{(\phi_p \varepsilon)^{1/4\alpha}} \qquad (50)$$

where $\alpha = 1/3$ holds for the Lifschitz–Slyozov process. Afterwards growth of domain size d in the course of time decreases dramatically. Since the time evolution in the early stage of phase decomposition is neglected, Eq. (50) represents a lower limit of t_c.

As previously mentioned, the quantity t_c is governed by inner parameters of the system. When one considers morphology evolution in polymer blend solutions, t_c has to be compared with the externally imposed pinning-down time t_p after which no further phase changes occur. For blend solutions, the quantity t_p is related to the rate of solvent evaporation.

According to Eq. (50) the characteristic time t_c is influenced by both the growth rate of domain size which is proportional to the inverse viscosity and the thermodynamic driving force represented by ε. Now, the conditions for the existence of regular phase morphologies in the late stage of phase decomposition can be specified more precisely as follows

– In the course of solvent induced phase separation periodic structures can be detected only when t_c exceeds t_p, $t_c - t_p > 0$.

– The "quench depth" ε must be sufficiently small, but, $\varepsilon \gg \dfrac{1}{r}$ is necessary for validity of the mean-field approximation.
– The growth rate of domain size which is proportional to $\dfrac{kT}{\eta}$ must be sufficiently small.

Morphology evolutions, both solvent induced and by thermal agitation, as sketched above can be qualitatively explained in terms of Eq. (50). Percolation limits for blends of PMMA and $S_x MMA_{1-x}$ were studied in Ref. [107]. Also these results agree quite well with the outlined principles.

4 Phase Behavior of Polymer Blends under Flow

4.1 Experimental Results

As discussed in Sect. 2, miscibility of polymers occurs mostly in a limited range of temperature and composition. Moreover, one may conclude that equilibrium phase behavior of polymer blends is ruled by a subtle balance of enthalpic and entropic contributions to the Gibbs free energy of mixing. In contrast, the phase behavior of blends under the conditions of flow is largely unexplored both experimentally and theoretically. It remains a challenging question how externally imposed forces affect the phase behavior of blends. This is important for the better understanding of non-equilibrium phase transitions as well as the utilization of flow induced phenomena in polymer blend processing.

Flow imparts both extension and rotation to fluid elements. Thus, polymer molecules will be oriented and stretched under these circumstances and this may result in flow-induced phenomena observed in polymer systems which include phase-changes, crystallization, gelation or fiber formation. More generally, the Gibbs free energy of polymer blends or solutions depends under non-equilibrium conditions not only on temperature, pressure and concentration but also on the conformation of the macromolecules (as an internal variable) and hence, it is sensitive to external forces.

Before discussing theoretical approaches let us review some experimental results on the influence of flow on the phase behavior of polymer solutions and blends. Pioneering work on shear-induced phase changes in polymer solutions was carried out by Silberberg and Kuhn [108] on a polymer mixture of polystyrene (PS) and ethyl cellulose dissolved in benzene; a system which displays UCST behavior. They observed shear-dependent depressions of the critical point of as much as 13 K under steady-state shear at rates up to 270 s^{-1}. Similar results on shear-induced homogenization were reported on a 50/50 blend solution of PS and poly(butadiene) (PB) with dioctyl phthalate (DOP) as a solvent under steady-state Couette flow [109, 110]. A semi-dilute solution of the mixture containing 3 wt% of total polymer was prepared. The quiescent solution exhibited a cloud point (an UCST) at 54 °C. Experiments under flow were performed at about 43 °C. The concentration fluctuations as influenced by shear flow were studied by small angle light scattering (SALS) perpendicular and parallel to the flow direction [110]. At a sufficiently high rate of shear ($\dot{\gamma} = 165\,\text{s}^{-1}$) it could be observed that the SALS intensity normal to flow direction is suppressed to the level of critical scattering in the quiescent solution. In other words, at $\dot{\gamma} = 165\,\text{s}^{-1}$ the critical temperature is equal to approximately 43 °C which corresponds to a quite dramatic drop of the critical temperature. This effect is indicative of the solution being transferred into the single phase region under the action of shear, wherefore the phenomenon is called shear-induced homogenization.

Qualitatively, the same effect has been observed in ternary solutions of p-PODZ, PA-6 and sulfuric acid [104]. At room temperature the quiescent system displays phase separation above 14% of total polymer concentration. Above the critical concentration shearing of initially biphasic solutions led to transparent one-phase systems. After cessation of the shear stress the biphasic morphologies recovered.

As mentioned above, the life-time τ of concentration fluctuations in polymer systems is very long. On the other hand, the distortion time of the concentration fluctuations by shear is $\dot{\gamma}^{-1}$ and one may establish that $\tau > \dot{\gamma}^{-1}$. As a result, the polymer system is highly sensitive to shear which results in distortion and suppression of concentration fluctuations by shear.

It should be emphasized that the effects produced by shear have been found different perpendicular and parallel to flow direction for the system PS/PB in DOP. The scattered intensity parallel to flow is weaker than that perpendicular to flow indicating that the fluctuations parallel to flow are much more suppressed by shear than those normal to flow. Similar results obtained by small angle neutron scattering for the same system were reported in Ref. [111].

The opposite behavior as sketched before was detected for solutions of PS in DOP [112]. Again, the critical temperature (an UCST at $T_c = 12\,°C$ in the quiescent state) turned out to be a function of the shear rate to which the solution is subjected. But, in contrast to solutions of PS and PB in DOP, here enhancements of the UCST as large as 28 °C were recorded at a shear rate of $220\,\text{s}^{-1}$. Similar results have been found for PS solutions in di(2-ethyl hexyl)phthalate or in a mixture of cis- and trans-decalin [113]. The solutions demixed in a converging flow from a reservoir into a capillary tube. It has been observed that an increase in the deformation rate raised the UCST or reduced the region of miscibility. In both of these studies an increase of the cloud point temperature of the polymer solutions was used as an indication of phase separation.

As before, one may establish that small changes in the deformation rate are accompanied by dramatic changes in the phase behavior. Furthermore, the first normal stress difference was reported in Ref. [112] as a function of polymer concentration at constant temperature and shearing stress. For the discussion below, it is important to keep in mind that the plots of the normal stress difference as a function of polymer concentration exhibit positive deviation from additivity for the systems under discussion. This result is in accord with other observations of dilute polymer solutions with upper critical solution temperatures which showed that phase separation is produced by mechanical deformation or, equivalently, that the critical temperature ascends with flow [112]. However, as mentioned before, it has been also reported on shear-dependent UCST depression which seems to be characteristic for solutions of polymer blends.

There are a number of studies of flow-induced phase changes in solutions, however, surprisingly few studies exist on the phase behavior of polymer blends subjected to flow. The primary observation has been flow-induced miscibility. Moreover, fluorescence evidence of shear-flow induced miscibility in blends of PS and poly(vinyl methyl ether) (PVME) was given in Ref. [114] and confirms that shear flow may induce mixing on a molecular scale. For these blends studied under steady shear flow using a cone-plate geometry an increase of the LCST by 2–7 K at shear rates of 0.1–20 s^{-1} was observed [115, 116]. Similar changes were reported for PMMA/SAN blends [117, 118]. For PMMA blended either with SAN-25 or SAN-29 (code: wt% AN in SAN) shear-induced miscibility was observed in a cone-plate geometry [117]. When the samples were quenched below the glass transition temperature after shearing no structure could be detected with transmission electron microscopy. Thus, it was possible to record shear-induced homogenization above LCST. After cessation of shear above LCST spinodal decomposition occurred. The same effect was detected for PMMA/SAN-31.5 blends subjected to steady-shear flow of Couette-type using a plate–plate geometry [118]. The LCST was elevated with increasing rate of shear. The shift amounted to 10 K at 4 s^{-1}. Again, the effect of shear turned out to be reversible.

Winter et al. [119, 120] studied phase changes in the system PS/PVME under planar extensional as well as shear flow. They developed a lubricated stagnation flow by the impingement of two rectangular jets in a specially built die having hyperbolic walls. Change of the turbidity of the blend was monitored at constant temperature. It has been found that flow-induced miscibility occurred after a duration of the order of seconds or minutes [119]. Miscibility was observed not only in planar extensional flow, but also near the die walls where the blend was subjected to shear flow. Moreover, the period of time required to induce miscibility was found to decrease with increasing flow rate. The LCST of PS/PVME was elevated in extensional flow as much as 12 K [120]. The shift depends on the extension rate, the strain and the blend composition. Flow-induced miscibility has been also found under shear flow between parallel plates when the samples were sheared near the equilibrium coexistence temperature. However, the effect of shear on polymer miscibility turned out to be less dramatic than the effect of extensional flow. The cloud point increased by 6 K at a shear rate of 2.9 s^{-1}.

Flow-induced miscibility was also found for blends of poly(ethylene-*co*-vinyl acetate) and solution chlorinated polyethylene undergoing simple shear flow at constant stress [121]. The LCST was shifted upward in temperature of about 25 K. Enhanced miscibility could also be found for the completely immiscible polymers SAN and polycarbonate in their blends when exposed to extremely high shear fields (up to $10^7 \, s^{-1}$) [122].

In Ref. [120] the first time has been reported on flow-induced phase separation in polymer blends. When PS/PVME blends were exposed to shear or extensional flow at lower temperatures, 20 to 30 K below the equilibrium coexistence temperature, phase separation was observed in both flow regimes. As the authors suggest, the stress, rather than the deformation rate, appears to be the most important parameter in flow-induced phase separation.

4.2 Theoretical Approaches

Several attempts have been made to explain theoretically the effects of flow on the phase behavior of polymer solutions [112, 115–118, 123, 124]. This has been done by modification of the mean-field free energy. The key point is to include properly the elastic energy of deformation produced by flow. A more rigorous approach originates from Helfand et al. [125, 126] and Onuki [127, 128] who proposed hydrodynamic theories for the dynamics of concentration fluctuations in the presence of flow coupled with a linear stability analysis.

Generally, one may establish that in some cases greatly enhanced concentration fluctuations occur under flow, in others, however, the size of concentration fluctuations is reduced and, obviously, flow promotes mutual miscibility of the polymers. Concentration fluctuations are accompanied by inhomogeneities of transport quantities as shear viscosity and diffusity. In a flow field the molecules are transferred into a non-equilibrium situation of extension. Two polymer molecules in a state of excess extension feel an additional repulsion due to the enhanced normal stress difference. Thus, the rate of dissipation by diffusion is low compared with the shear rate and the concentration fluctuations tend to grow. The opposite is true for a state of lower extension. In that case the dissipation of the concentration fluctuations is enhanced owing to an additional attraction between the chain molecules.

In the following, the discussion will be restricted to flow-induced phase changes in polymer blends under steady-state conditions. Then, a quasi-thermodyamic approach is certainly justified when

$$\tau > \dot{\gamma}^{-1} \tag{51}$$

where τ and $\dot{\gamma}$ represent the relaxation time of concentration fluctuations in the blend and the rate of shear, respectively. In the limit indicated by Eq. (51) the explanation of shear-dependent phase behavior is straightforward. The system exposed to shear can store energy, i.e. an additional contribution to the free energy results. Hence, the phase behavior must be affected by shear. When shear is stopped the original equilibrium state is restored after a while. Therefore, the effect resulting from shear stress must be – in thermodynamic terms – an elastic free energy contribution.

In the simplest approximation one may start with the free energy change of a single chain under steady-state flow. It reads

$$\Delta g_{el} = \frac{3}{2} kT \left[\frac{R^2}{R_0^2} - 1 \right] \tag{52}$$

where R^2 and R_0^2 are the mean square of the end-to-end distance of the flowing system and at rest, respectively. If one assumes further that Eq. (52) holds good also for a system of N chains then the additional term to the molar Gibbs free energy caused by flow under steady-state conditions can be related to the trace of the deviatoric stress tensor P which is a measure of the molecular deformation and which equals the first normal stress difference N_1 under steady laminar shearing flow [129, 130]. It follows

$$G_{el} = V \tfrac{1}{2} TrP = \tfrac{1}{2} VN_1 \tag{53}$$

where V is the molar volume of the system. It is important to note that the free energy change of a macroscopic system under flow, as given by Eq. (53), is reduced here to the deformation behavior of individual polymer molecules. Effects of intermolecular interactions on the deformation are neglected. In that respect, Eq. (53) is similar to the well-known expressions for rubber elasticity [131].

For a polymer mixture one has to take into account that the normal stress difference is a function of composition. Thus, in generalization of Eq. (53) it is appropriate to assume that N_1 of the mixture follows the simple but adequate relationship

$$N_1 = N_{1A} \phi_A + N_{1B} \phi_B + \Delta N_1 \phi_A \phi_B \tag{54}$$

where N_{1i} and ϕ_i represent the normal stress difference and the volume fraction of pure component i, respectively, ΔN_1 describes the excess normal stress difference of the mixture (the deviation from additivity) and depends on the rate of shear. Equations (53) and (54) are suitable for calculating the extra elastic free energy of mixing caused by shear. Introducing in analogy to Sect. 2, Eq. (6), reduced state variables, it follows

$$\frac{\Delta G_{el}^M}{RT} = \frac{1}{2} \frac{V_A^* P_A^*}{RT_A^*} \frac{\tilde{V}_A}{\tilde{T}_A} \cdot \Delta \tilde{N}_1 \phi_A \phi_B \tag{55}$$

Here, $\tilde{N}_1 \equiv N_1/P^*$ is the reduced normal stress difference. Employing Eq. (AIV 1) for the molar configurational energy U_A of component A one may recast Eq. (55) into

$$\frac{\Delta G_{el}^M}{RT} = -\frac{1}{2} \frac{U_A}{RT} \tilde{V}_A^2 \Delta \tilde{N}_1 \phi_A \phi_B \tag{56}$$

Now, we assume that expression (56) can be simply added to the Gibbs free energy of mixing (Eqs. (1)–(3)). Strictly speaking, that is an additional approximation which implies that the free energy of mixing (Eq. (1)) is not affected additionally by the polymer coil deformation leading to the stored elastic energy of Eq. (56). Thus, in the context of the approximations mentioned, the effect of flow can be expressed by an additional contribution to the free-energy parameter X, Eq. (14). Neglecting, for the sake of simplicity, the size-effect, $\rho = 0$, instead of Eq. (14) we obtain

$$X = -\frac{U_A}{RT}\left(2X_{AB} + \frac{1}{2}\tilde{V}_A^2\Delta\tilde{N}_1\right) + \frac{C_{VA}}{R}\frac{7}{8}\Gamma^2 \tag{57}$$

As mentioned before, for high-molar-mass polymers to be miscible the free-energy parameter X must be negative which requires $X_{AB} < 0$. In the approximation represented by Eq. (57), flow corresponds to an additional (shear-rate dependent) interaction, but, it does not – in that approximation – cause any change of the free-volume term. For negative excess elastic energy, i.e. $\Delta N_1 < 0$, a polymer blend exposed to shear feels a stronger interaction between the different segments leading to enhanced miscibility of the two components. This shear homogenization originates from an additional ordering in the system leading to favorable interchain interactions which for an LCST are counterbalanced by a corresponding entropy reduction.

Let us add here some remarks on the normal stress difference. According to the Rouse–Zimm model [132, 133] the first normal stress difference may be related to the storage modulus G'. Taking into account only the longest relaxation time τ_1 one gets

$$G' = G_N^0\frac{(\dot{\gamma}\tau_1)^2}{1+(\dot{\gamma}\tau_1)^2} \qquad N_1 = 2G_N^0\cdot(\dot{\gamma}\tau_1)^2 \tag{58}$$

where G_N^0 is the plateau modulus. The relation between N_1 and G' depends on the magnitude of the quantity $\dot{\gamma}\tau_1$:

$$\begin{array}{ll} N_1 = 4G' = 2G_N^0 & \dot{\gamma}\tau_1 = 0(1)^{*)} \\ N_1 = 2G' & \text{for} \qquad \dot{\gamma}\tau_1 \ll 1 \end{array} \tag{59}$$

As can be seen, determination of the plateau modulus as a function of blend composition yields ΔG_N^0, which is closely related to ΔN_1. Thus, the free-energy change under steady-state shear is readily accessible experimentally.

Moreover, Wu related ΔG_N^0 for miscible polymers to entanglement density [134]. For $\Delta G_N^0 < 0$ the entanglement density in the miscible system is reduced compared to additivity behavior which means that the different chain molecules tend to align in the mixture. This is the same effect as stated above. The shift of the spinodal temperature caused by flow can be estimated easily as follows. The spinodal may be given as in the case of a system at rest

$$\left(\frac{\partial^2\Delta G^M}{\partial\phi^2}\right)_{P,T,\dot{\gamma}} \equiv \Delta G^{M''} = 0 \tag{60}$$

where ΔG^M now contains also the elastic contribution (56).

---

*) The symbol "0" is for "order of magnitude".

Expanding $\Delta G^{M''}$ around the spinodal temperature T_0 for the quiescent blend, it follows that

$$\Delta G^{M''}(T) = \Delta G_0^{M''}(T_0) + \Delta G_{el}^{M''}(T_0) + (T - T_0)\left(\frac{\partial \Delta G^{M''}}{\partial T}\right)_{T_0} = 0 \qquad (61)$$

where all quantities refering to the blend at rest are labeled by the index o. The first term on the right-hand side of Eq. (61) marks the stability limit of the quiescent blend, hence, it vanishes. The derivative in the last term equals the respective derivative of the entropy of mixing, $-\Delta S^{M''}(T_0)$. Putting $\Delta T \equiv (T - T_0)$ one gets for the temperature shift

$$\Delta T = \frac{\Delta G_{el}^{M''}(T_0)}{\Delta S^{M''}(T_0)} \qquad (62)$$

Using Eqs. (55) and (20) one arrives at

$$\Delta T = -\frac{V}{2RT_0}\frac{\Delta N_1}{\left(\dfrac{\partial X}{\partial T}\right)_{T_0}} \qquad (63)$$

As outlined in Sect. 2 for miscible polymers with an LCST, the parameter X increases with temperature, hence, $\left(\dfrac{\partial X}{\partial T}\right) > 0$. For $\Delta N_1 < 0$ it results in $\Delta T > 0$. One sees, for an LCST the shear homogenization is directed by the balance of a negative elastic energy contribution and a negative entropy term to the Gibbs free energy of mixing. The opposite is true for a UCST owing to $\left(\dfrac{\partial X}{\partial T}\right)_{UCST} < 0$. This gives

$$\Delta T \gtrless 0 \cdot \begin{matrix} LCST \\ UCST \end{matrix} \quad \text{if} \quad \Delta N_1 < 0$$

For the blend PMMA/SAN-31.5 (code: 31.5 wt% of AN in SAN) a phase separation temperature of 184 °C has been observed for the quiescent 60/40 blend which shifted by about 10 K at a shear rate of $2\,s^{-1}$ [118]. All the parameters occurring in Eq. (63) are known for the system under discussion except the quantity ΔN_1. Thus, ΔN_1 can be estimated. Employing the individual segmental interaction parameters submitted in Ref. [6] one easily arrives at

$$X_{AB} = -9.6 \times 10^{-4} \qquad \Gamma^2 = 1.5 \times 10^{-4}$$

for the system PMMA/SAN-31.5. Moreover, applying Eq. (15′) with $T_{PMMA}^* = 8100$ K [6] the phase separation temperature of the quiescent blend corresponds to $\tilde{V}_A^{1/3} = 1.0755$. Inserting all these values in Eq. (63), it follows immediately $\Delta \tilde{N}_1 = -7 \times 10^{-5}$. An adequate value for the reduced pressure is $P_A^* = 5 \times 10^8$ Pa [135]. Therefrom, one gets for the specific elastic excess energy: $\Delta N_1 = -3.5 \times 10^4$ Pa. Precise measurements of the plateau modulus as a function of composition for the system PMMA/SAN-24.1 at 180 °C were submitted in Ref. [134]. From this result one can extract $\Delta G_N^0 = -0.2$ MPa which leads to $\Delta N_1 = -4 \times 10^5$ Pa according to Eq. (59).

The estimated value and the experimental value for ΔN_1 differ by one order of magnitude. However, it is evident that the theory yields correctly the sign of the excess elastic energy contribution caused by flow. In other words, the theory can predict the direction in which a UCST or LCST is shifted under the action of flow. In conclusion, the phenomenological approach can explain, at least qualitatively, flow-induced phase changes in polymer blends.

According to Eq. (63) and the estimation of the quantity ΔN_1 reported above the temperature shift in LCST caused by flow should be one order of magnitude higher than observed. This discrepancy may result from the fact that the elastic free energy of Eq. (53) is regarded as the sum of the energy changes in the individual chain molecules. Thus, this term is determined exclusively by intramolecular effects, intermolecular interactions between chains are considered as constant during deformation. As a result, the free-volume term is not affected by flow in that approximation, cf. Eq. (57). In other words, the volume change in a blend undergoing phase separation under flow and the accompanying change of intermolecular interactions is not adequately included in the presented model.

From a theoretical point of view the most basic open question is whether the coexistence curve under flow can be understood in the context of a properly generalized thermodynamic theory or whether it is necessary to use kinetic theories. We tend to the opinion that thermodynamic approaches can provide valuable insight into the observed effects.

5 Appendices

Appendix I. The parameters X_{AB}, Γ, ρ and K for blends containing a random copolymer.

As mentioned above for copolymer based blends the quantities ε_{ij}^* and r_{ij}^* in definitions (9) have to be replaced by number-averaged quantities $\langle \varepsilon_{ij}^* \rangle$ and $\langle r_{ij}^* \rangle$. For a blend of a homopolymer A and a random copolymer B, $P(C_\beta D_{1-\beta})$, comprising segments of type C and D with mole fractions β and $(1 - \beta)$, respectively, the generalization of the parameters can be done along the same lines as sketched in Appendix III. It is useful to introduce the following notations:

$$\delta_i \equiv \frac{\varepsilon_{ii}^*}{\varepsilon_{AA}^*} - 1 \qquad \delta_i^r \equiv \frac{r_{ii}^*}{r_{AA}^*} - 1 \quad i = C, D$$

$$\chi_{ij} \equiv \frac{1}{\varepsilon_{AA}^*} \left[\frac{1}{2}(\varepsilon_{ii}^* + \varepsilon_{jj}^*) - \varepsilon_{ij}^* \right]$$

$$R_{ij} \equiv \frac{1}{r_{AA}^*} \left[\frac{1}{2}(r_{ii}^* + r_{jj}^*) - r_{ij}^* \right] \qquad i \neq j; \quad \begin{matrix} i = A, C \\ j = C, D \end{matrix}$$

The generalized parameter R_{AB} for copolymers of Eq. (9) will be represented by

K. Then, for the parameters one obtains

$$X_{AB} = \beta\chi_{AC} + (1 - \beta)\chi_{AD} - \beta(1 - \beta)\chi_{CD} \tag{I1}$$

$$\Gamma = \beta\delta_C + (1 - \beta)\delta_D - 2\beta(1 - \beta)\chi_{CD} - 18\beta(1 - \beta)(\delta_C^r + \delta_D^r)^2 \tag{I2}$$

$$K = -\beta R_{AC} - (1 - \beta)R_{AD} + \beta(1 - \beta)R_{CD} \tag{I3}$$

$$\rho = \beta\delta_C^r + (1 - \beta)\delta_D^r \tag{I4}$$

All the relations listed in Appendix III can be used with the parameters as given in Eqs. (I1)–(I4).

Appendix II. The derivatives of parameter X with respect to the reduced temperature and volume – Eq. (19).

$$\left(\frac{\partial X}{\partial \frac{1}{\tilde{T}}}\right)_A = \tilde{T}_A \frac{U_A}{RT} \qquad \left(\frac{\partial X}{\partial \left(\frac{1}{\tilde{T}}\right)^2}\right)_A = -\tilde{T}_A^2 \frac{C_{VA}}{R}$$

$$\left(\frac{\partial}{\partial \left(\frac{1}{\tilde{T}}\right)} \frac{\partial X}{\partial \tilde{V}}\right)_A = \tilde{T}_A \left(\frac{\partial \tilde{P}}{\partial \tilde{T}}\right)_A \qquad \left(\frac{\partial^2 X}{\partial \tilde{V}^2}\right)_A = \frac{1}{\tilde{T}_A \tilde{V} \tilde{\kappa}_A} = -\frac{U_A}{RT} \frac{1}{\tilde{\kappa}_A}$$

Appendix III. Explicit expressions for the quantities θ_i and Ω_i.

Employing a (6–12) law for the potential function $\varepsilon(r')$ Eq. (7) the parameters $\langle \varepsilon_A^* \rangle$ and $\langle r_A^* \rangle$ are given for a pair of homopolymers by

$$\langle \varepsilon_A^* \rangle = \frac{(\phi_A \varepsilon_{AA}^* r_{AA}^{*6} + \phi_B \varepsilon_{AB}^* r_{AB}^{*6})^2}{\phi_A \varepsilon_{AA}^* r_{AA}^{*12} + \phi_B \varepsilon_{AB}^* r_{AB}^{*12}} \tag{III1}$$

$$\langle r_A^* \rangle = \left(\frac{\phi_A r_{AA}^{*12} + \phi_B r_{AB}^{*12}}{\phi_A r_{AA}^{*6} + \phi_B r_{AB}^{*6}}\right)^{1/6}$$

Similar expressions exist for $\langle \varepsilon_B^* \rangle$ and $\langle r_B^* \rangle$. From Eq. (III1) combined with the definitions (9) one obtains

$$\theta_A = \frac{\phi_B}{2} \Gamma - \phi_B X_{AB} - 9 \rho^2 \phi_A \phi_B \tag{III2}$$

$$\theta_B = \left(\phi_B + \frac{\phi_A}{2}\right) \Gamma - \phi_A X_{AB} - 9 \rho^2 \phi_A \phi_B \tag{III3}$$

and

$$\frac{\langle r_A^* \rangle}{r_{AA}^*} = 1 + \frac{\rho}{2} \phi_B + K\phi_B \tag{III4}$$

$$\frac{\langle r_B^* \rangle}{r_{AA}^*} = 1 + \rho\left(\frac{\phi_A}{2} + \phi_B\right) + K\phi_A \tag{III5}$$

where $K = -R_{AB}$.

For the mixture an equation-of-state exists:

$$\tilde{P} = \tilde{P}(\langle \tilde{V} \rangle, \langle \tilde{T} \rangle)$$

Expanding this equation-of-state in terms of pure component A at the limit $\tilde{P} = 0$ one can deduce the quantities Ω as follows

$$\tilde{P} = 0 = \left(\frac{\partial \tilde{P}}{\partial \frac{1}{\tilde{T}}} \right)_A (\phi_A \theta_A + \phi_B \theta_B) \cdot \frac{1}{\tilde{T}_A}$$

$$+ \left(\frac{\partial \tilde{P}}{\partial \tilde{V}} \right)_A [\Omega_A + \phi_B(\langle \tilde{V}_B \rangle - \langle \tilde{V}_A \rangle)]$$

Now, let $\langle \tilde{V}_B \rangle = \tilde{V}_A + (\langle \tilde{V}_B \rangle - \tilde{V}_A)$ then

$$\Omega_A = \frac{\left(\tilde{T} \frac{\partial \tilde{P}}{\partial \tilde{T}} \right)_A}{\left(\frac{\partial \tilde{P}}{\partial \tilde{V}} \right)_A} \sum_i^{A,B} \phi_i \theta_i + \phi_B \tilde{V}_A \left(\frac{\langle r_B^* \rangle^3}{\langle r_A^* \rangle^3} - 1 \right) \tag{III6}$$

and an analogous expression for Ω_B where in the second term ϕ_B is replaced by $-\phi_A$. For Ω_{BB} the same calculation yields

$$\Omega_{BB} = \frac{\left(\tilde{T} \frac{\partial \tilde{P}}{\partial \tilde{T}} \right)_A}{\left(\frac{\partial \tilde{P}}{\partial \tilde{V}} \right)_A} \theta_{BB} \tag{III7}$$

Appendix IV. Prefactors of Eq. (14) expressed by reduced quantities using Flory's equation-of-state (15)

$$-\frac{U_A}{RT} = \frac{1}{\tilde{T}_A \tilde{V}_A} \tag{IV1}$$

$$-\frac{\tilde{V}_A^2 U_A}{RT \tilde{\kappa}_A} = \frac{\frac{4}{3} - \tilde{V}_A^{1/3}}{\tilde{T}_A \tilde{V}_A (\tilde{V}_A^{1/3} - 1)} \tag{IV2}$$

$$\frac{C_{VA}}{R} = \frac{\tilde{V}_A^{1/3}}{\frac{4}{3} - \tilde{V}_A^{1/3}} \tag{IV3}$$

All expressions at the right-hand side have to be multiplied by $P_A^* V_A^*/RT_A^*$.

6 Abbreviations for Polymers

EPR	ethylene-propylene copolymer
EVA	poly(ethylene-co-vinylacetate)
MMA-BMA	poly(methyl methacrylate-co-butyl methacrylate)
MSAN	poly(α-methyl styrene-co-acrylonitrile)
PA	polyamide
PB	poly(1,4-cis-butadiene)
PBZT	poly(p-phenylene benzobisthiazole)
PCL	poly(ϵ-caprolactone)
PE	polyethylene
PEA	poly(ethyl acrylate)
PETP	poly(ethylene terephthalate)
PMMA	poly(methyl methacrylate)
PODZ	poly(p-phenylene 1,3,4-oxadiazole)
PP	polypropylene
PPO	poly(2,6-dimethyl phenylene oxide)
PPTA	poly(p-phenylene terephthalamide)
PS	polystyrene
PVC	poly(vinyl chloride)
PVDF	poly(vinylidene fluoride)
PVME	poly(vinyl methyl ether)
SAN	poly(styrene-co-acrylonitrile)
SB	poly(styrene-co-butadiene)
SMA	poly(styrene-co-maleic anhydride)
SMMA	poly(styrene-co-methyl methacrylate)

7 References

1. Kambour RP, Bendler JT, Bopp RC (1983) Macromolecules 16: 753
2. tenBrinke G, Karasz FE (1984) Macromolecules 17: 815
3. Kammer HW (1986) Acta Polymerica 37: 1
4. Karasz FE, MacKnight WJ (1986) Adv Chem Ser 211: 67
5. Lath D, Cowie JMG (1988) Makromol Chem Macromol Symp 16: 103
6. Kammer HW, Kressler J, Kressler B, Scheller D, Kroschwitz H, Schmidt-Naake G (1989) Acta Polymerica 40: 75
7. Litauszki B, Schmidt-Naake G, Kressler J, Kammer HW (1989) Polym Commun 30: 359
8. Kammer HW, Inoue T, Ougizawa T (1989) Polymer 30: 888
9. Ougizawa T, Inoue T, Kammer HW (1985) Macromolecules 18: 2089
10. Ougizawa T, Inoue T (1986) Polymer J (Tokyo) 18: 521
11. Saito H, Fujita Y, Inoue T (1987) Polymer J (Tokyo) 19: 405
12. Cong G, Huang Y, MacKnight WJ, Karasz FE (1986) Macromolecules 19: 2765
13. Hashimoto T (1988) In: Komura S, Furukawa H (eds) Dynamics of ordering. Plenum, New York, p 421

Hashimoto T (1987) In: Ottenbrite RM, Utracki LA, Inoue S (eds) Current topics in polymer science. Hauser, New York, p 199
14. Nose T (1987) Phase Transitions 8: 245
15. Izumitani T, Hashimoto T (1985) J Chem Phys 83: 3964
16. Binder K (1987) Colloid & Polymer Sci 265: 273
17. Strobl GR (1985) Macromolecules 18: 558
18. Kammer HW, Pigłowski J (1988) Z Phys Chem (Leipzig) 269: 721
19. Inaba N, Sato K, Suzuki S, Hashimoto T (1986) Macromolecules 19: 1690
20. Inoue T, Ougizawa T (1989) J Macromol Sci-Chem A26: 147
21. Kammer HW, Kressler J, Kummerlöwe C, Morgenstern B (1990) Polimery (Poland) 35: 199
22. Flory PJ (1970) Disc Faraday Soc 49: 7
23. Tsujita Y, Iwakiri K, Kinoshita T, Takizawa A, MacKnight WJ (1988) J Polym Sci-Polym Phys B25: 415
24. Li Y, Wold M, Wendorff JH (1987) Polym Commun 28: 265
25. Kammer HW (1991) Polymer 32: 501
26. Brandrup J, Immergut EH (1966) Polymer handbook. Interscience, New York
27. tenBrinke G, Eshius A, Roerdink E, Challa G (1981) Macromolecules 14: 867
28. Riedl B, Prud'homme RE (1988) J Polym Sci-Polym Phys B26: 1769
29. tenBrinke G, Karasz FE, MacKnight WJ (1983) Macromolecules 16: 1827
30. Goh SH, Lee SY (1988) Thermochim Acta 123: 3
31. Kressler J, Kammer HW, Morgenstern U, Litauszki B, Berger W, Karasz FE (1990) Makromol Chem 191: 243
32. Kressler J, Kammer HW (1987) Acta Polymerica 38: 600
33. Goh SH, Lee SY (1987) Eur Polym J 23: 315
34. Goh SH, Lee SY (1989) Eur Polym J 25: 997
35. Goh SH, Lee SY (1990) Eur Polym J 26: 711
36. Chien YY, Pearce EM, Kwei TK (1988) Macromolecules 21: 1616
37. Min KE, Paul DR (1988) J Polym Sci-Polym Phys B26: 2257
38. Cowie JMG, McEwen IJ, Nadvornik L (1990) Macromolecules 23: 5106
39. Bierlich M, Kressler J, Kammer HW (1991) Trends Polym Sci (India) 2: 9
40. Alexandrovich P, Karasz FE, MacKnight WJ (1977) Polymer 18: 1022
41. Suess M, Kressler J, Kammer HW (1987) Polymer 28: 957
42. Stein DJ, Jung RH, Illers KH, Hendus H (1974) Angew Makromol Chem 36: 89
43. Fowler ME, Barlow JW, Paul DR (1987) Polymer 28: 1177
44. Suess M, Kressler J, Kammer HW, Heinemann K (1986) Polym Bull 16: 371
45. Goh SH, Lee SY (1990) Eur Polym J 26: 715
46. Vukovic R, Bogdanovic G, Kuresevic V, Karasz FE, MacKnight WJ (1988) Eur Polym J 24: 123
47. Chiu SC, Smith TGJ (984) Appl Polym Sci 29: 1797
48. Kressler J, Kammer HW, Herzog K, Heyde H (1990) Acta Polymerica 41: 1
49. Pigłowski J (1988) Eur Polym J 24: 905
50. Cantow HJ, Schulz D (1986) Polym Bull 15: 449
51. Balasz AC, Karasz FE, MacKnight WJ, Ueda H, Sanchez IC (1985) Macromolecules 18: 2784
52. Hellmann GP, Kohl PR, Herth J, Neumann HJ, Andradi LN, Löwenhaupt B (1990) Makromol Chem Macromol Symp 38: 17
53. Zhikuan C, Ruona S, Karasz FE (1992) Macromolecules
54. Cowie JMG, Reid VMC, McEwen IJ (1990) Polymer 31: 905
55. Kang HS, MacKnight WJ, Karasz FE (1987) Polym Prepr (Am Chem Soc Div Polym Chem.) 28: 134
56. Ruona S, Zhikuan C, Karasz FE (1990) Macromolecules 23: 5
57. Shiomi T, Karasz FE, MacKnight WJ (1986) Macromolecules 19: 2274
58. Hall WJ, Cruse RL, Mendelson RA, Trementozzi QA (1982) Am Chem Soc Div Org Coat Plast Chem 47: 298
59. Kim JH, Barlow JW, Paul DR (1989) J Polym Sci-Polym Phys B27: 223
60. Kressler J, Kammer HW, Schmidt-Naake G, Herzog K (1988) Polymer 29: 686
61. Aoki Y (1988) Macromolecules 21: 1277
62. Cowie JMG, Reid VMC, McEwen IJ (1990) Polymer 31: 486
63. Kammer HW, Pigłowski J (1989) Acta Polymerica 40: 363
64. Nose T (1989) In: Tanaka F, Doi M, Ohta T (eds) Space-time organization in macromolecular fluids. Springer, Berlin Heidelberg New York, p 40 (Series in Chemical Physics, vol 51)

65. Cahn JW (1961) Acta Met 9: 795
    Cahn JW (1965) J Chem Phys 42: 93
66. Vrij A, van der Esker MWJ (1972) J Chem Soc Faraday Trans II68: 513
67. Nose T (1976) Polym J (Tokyo) 8: 96
68. Higgins JS, Fruitwala H, Tomlins PE (1988) Makromol Chem Macromol Symp 16: 313
69. Kammer HW, Kressler J (1988) Makromol Chem Macromol Symp 18: 63
70. Okada M, Han CC (1986) J Chem Phys 85: 5317
71. Snyder HL, Meakin P, Reich S (1983) Macromolecules 16: 757
72. Meier H, Strobl GR (1987) Macromolecules 20: 649
73. Hill RG, Tomlins PE, Higgins JS (1985) Macromolecules 18: 2556
74. de Gennes PG (1980) J Chem Phys 72: 4576
75. Binder K (1983) J Chem Phys 79: 6387
76. Hashimoto T, Itakura M, Hasegawa H (1986) J Chem Phys 85: 6773
77. Sasaki K, Hashimoto T (1984) Macromolecules 17: 2818
78. Langer JS, Bar-on M, Miller HD (1975) Phys Rev A11: 1417
79. Binder K, Stauffer D (1974) Phys Rev Lett 33: 1006
80. Kawasaki K (1977) Progr Theor Phys 57: 826
81. Kawasaki K, Ohta T (1978) Progr Theor Phys 59: 362
82. Furukawa H (1985) Adv Phys 34: 703
    Furukawa H (1984) Physica A123: 497
83. Furukawa H (1978) Progr Theor Phys 59: 1027
    Furukawa H (1981) Phys Rev A23: 1535
84. Lifschitz IM, Slyozov VV (1961) J Phys Chem Solids 19: 35
85. Siggia ED (1979) Phys Rev A20: 595
86. Bates FS, Wiltzius P (1989) J Chem Phys 91: 3258
87. Pigłowski J, Kressler J, Kammer HW (1986) Polym Bull 16: 493
88. Takagi Y, Ougizawa T, Inoue T (1987) Polymer 28: 103
89. Sato T, Han CC (1988) J Chem Phys 88: 2057
90. Kumaki J, Hashimoto T (1986) Macromolecules 19: 763
91. Kawahara S, Akiyama S (1990) Polymer J (Tokyo) 22: 361
92. Pigłowski J, Kammer HW, Kressler J (1989) Polymer 30: 1705
93. Inoue T, Ougizawa T, Yasuda O, Miyasaka K (1985) Macromolecules 18: 57
94. Morgenstern B, Kammer HW (1989) Polym Bull 22: 265
95. Schoene A, Morgenstern B, Berger W, Kammer HW (1991) Polymer Networks & Blends 1: 109
96. Reichelt K, Kummerloewe C, Kammer HW (1992) Acta Polymerica 43: 17
97. Inaba N, Yamada T, Suzuki S, Hashimoto T (1988) Macromolecules 21: 407
98. Nojima S, Satoh K, Ashida T (1991) Macromolecules 24: 942
99. Nakai A, Shiwaku T, Hasegawa H, Hashimoto T (1986) Macromolecules 19: 3008
100. Nagaya T, Orihara H, Ishibashi Y (1989) J Phys Soc Jpn 58: 3600
101. Chuah HH, Kyu T, Helminiak TE (1989) Polymer 30: 1591
102. Kyu T, Chen TI, Park HS, White JL (1989) J Appl Polym Sci 37: 201
103. Kammer HW, Kummerloewe C (1990) Acta Polymerica 41: 269
104. Kummerloewe C, Kammer HW, Malinconico M, Martuscelli E (1991) Polymer 32: 2505
105. Helfand E, Tagami Y (1971) J Polym Sci B9: 741
106. Kammer HW (1977) Z Phys Chem (Leipzig) 258: 1149
107. Andradi LN, Hellmann GP (1992) Polymer
108. Silberberg A, Kuhn W (1952) Nature.170: 450
109. Wolf BA (1980) Makromol Chem Rapid Commun 1: 231
110. Takebe T, Hashimoto T (1988) Polym Commun 29: 227
111. Nakatani AI, Kim H, Takahashi Y, Han CC (1989) Polym Commun 30: 143
112. Rangel-Nafaile C, Metzner AB, Wissbrun K (1984) Macromolecules 17: 1187
113. VerStrate G, Phillipoff W (1974) J Polym Sci Lett 12: 267
114. Cheikh Larbi FB, Malone MF, Winter HH, Halary JL, Leviet MH, Monnerie L (1988)
     Macromolecules 21: 3532
115. Mazich KA, Carr SH (1983) J App Phys 54: 5511
116. Rector LP, Mazich KA, Carr SH (1988) J Macromol Sci-Phys B27: 421
117. Lyngaae-Jørgensen J, Søndergaard K (1987) Polym Eng Sci 27: 344, 351
118. Kammer HW, Kummerloewe C, Kressler J, Melior JP (1991) Polymer 32: 1488
119. Katsaros JD, Malone MF, Winter HH (1986) Polym Bull 16: 83

120. Katsaros JD, Malone MF, Winter HH (1989) Polym Eng Sci 29: 1434
121. Hindawi I, Higgins JS, Galambos AF, Weiss RA (1990) Macromolecules 23: 670
122. Takahashi H, Matsuoka T, Ohta T, Fukumori K, Karauchi T, Kamigaito O (1988) J Appl Polym Sci 36: 1821
123. Vrahopoulou-Gilbert E, McHugh AJ (1984) Macromolecules 17: 2567
124. Romig KD, Hanley HJM (1986) Int J Thermophys 7: 877
125. Bhattacharjee SM, Frederickson GH, Helfand E (1989) J Chem Phys 90: 3305
126. Helfand E, Frederickson GH (1989) Phys Rev Lett 62: 2468
127. Onuki A (1989) Phys Rev Lett 62: 2472
128. Onuki A (1990) J Phys Soc Jpn 59: 3423, 3427
129. Janeschitz-Kriegl H (1969) Adv Polym Sci 6: 170
130. Marucci G (1972) Trans Soc Rheol 16: 321
131. Treloar LRG (1975) The physics of rubber elasticity, 3rd edn. Clarendon Press, Oxford, chap 5
132. Rouse PE (1953) J Chem Phys 21: 1273
133. Zimm B (1956) J Chem Phys 24: 269
134. Wu S (1987) Polymer 28: 1144
135. Patterson D, Robard A (1978) Macromolecules 11: 690

Editor: H.-H. Kausch
Received January 1992

SANS from Homogeneous Polymer Mixtures: A Unified Overview

Boualem Hammouda
Materials Science and Engineering Laboratory, National Institute of Standards and Technology, Building 235, Room E151, Gaithersburg, MD 2089, USA

An overview of various modeling methods used to understand small angle neutron scattering (SANS) data from homogeneous polymer systems is presented. First, calculations of single macromolecule structure factors are reviewed for many chain architectures and monomer block configurations such as linear, ring, star branched, comb grafted chains and regular "starburst" dendrimers either in the homopolymer or copolymer forms. Then, the different methods used to model "concentration" effects in polymer solutions (dilute, semidilute, concentrated), polymer melts and blend mixtures are summarized on the basis of the random phase approximation. Polymer chain stiffness is also included in the formalism so that mixtures of liquid crystals and flexible polymers in the single-phase region can be described. Specific examples are included along with various SANS data that were analyzed within this framework. This overview is meant to be a guide to help build up models in order to understand SANS data from many homogeneous polymer systems. It is not meant to be complete and is not an exhaustive review of the literature in the field. Most of the results discussed have been previously published and are brought together here in a unified self-contained approach.

1 Introduction

Since its introduction in the early 1970s, the small angle neutron scattering (SANS) method has had a substantial impact on polymer research. When used with partially deuterated polymers, SANS permits a close monitoring of macromolecular conformations in polymer solutions, melts, and blend mixtures. This advantage has made it a unique tool for the understanding of the morphology of polymer materials and of the relationship between their structures and physical properties.

Macromolecular systems can be modeled fairly well owing to the pioneering work of many scientists such as P. Debye (flexible chains, etc.), P. Flory (gaussian chains, theta temperature, etc.), H. Kuhn (polymer chain stiffness, etc.), W. Stockmayer (gelation, branching, etc.), B. Zimm (dilute solutions, normal modes, etc.), P.G. de Gennes (random phase approximation, scaling ideas, etc.), H. Yamakawa (wormlike chains, etc.), H. Benoit (star polymers. multicomponent description, etc.), A.Z. Akcasu (high concentration method, first cumulant, etc.), K.F. Freed (renormalization group theory) to name only a few. The SANS technique has shown with no ambiguity, for instance, that polymer coils form random walk trajectories when they are in the melt or bulk states [1]. This means that correlations between monomers along the chain backbone are screened by other surrounding chains, so that the chain "forgets" quickly (after one step) about where its other parts are [2]. SANS has been most valuable for the monitoring of chain conformations in a wide variety of polymer systems (solutions, melts, solids) and with a wide array of experimental conditions. It has also proven its usefulness, during the last few years, as a thermodynamic probe used to map out phase diagrams of polymer blend mixtures.

Blending of polymers is necessary for better controlled physical properties of polymeric materials. Unfortunately, most polymer blends are immiscible. The binary mixtures that are known to be miscible have been very valued systems for studying the thermodynamics of phase separation. Conformations in the miscible region, concentration fluctuations close to the immiscible region as well as the delimitation of the spinodal line have been well understood for a number of polymer blend systems using the SANS technique with deuterium labeling of one of the components. Our description of homopolymer and/or copolymer blends stops at the spinodal line in the sense that heterogeneous multiphase systems fall outside of the scope of this overview which is devoted to homogeneous single-phase mixtures only.

Advances in the modeling of polymer systems, such as the random phase approximation (RPA) [3, 4], thoroughly reviewed here, have made it possible to analyze SANS data from widely different and seemingly complicated mixtures of various polymer architectures at various concentrations and temperatures. Old modeling methods, such as the inverse Zimm formula [5] (which is the basis of the Zimm Plot), or more recent modeling methods such as the high concentration method [6–8], fall within the scope of the present overview and are discussed in detail since they constitute major tools for the understanding of SANS data from polymer solutions. Some misgivings as to the validity of using such "mean field" approaches to describe polymer solutions (where concentration fluctuations are important) have been presented in the literature [4, 9, 10]. A generalization of de Gennes' RPA formula [3, 4] based on a direct computations method [11, 12] or on a multicomponent matrix approach [13–15] are also reviewed and reproduced in an appendix in order to clarify what assumptions are made for its derivation. Another calculation [16] applied the RPA to ternary polymer blends. A further generalization of the multicomponent RPA method to include polymer chain stiffness is included. These recent results will be useful for the qualitative treatment of mixtures of liquid crystals and flexible polymers in the single-phase region. They can predict the isotropic-to-nematic phase transition as well as the spinodal line.

On the other hand, advances in chemical synthesis have made possible the polymerization of many complicated chain architectures. Linear, ring, star branched and comb grafted chains can be made now with high regularity, monodispersity and controlled sequencing of different chemical blocks. For instance, stars can be synthesized with a fraction (say one third) of one arm deuterated, copolymer combs can be made with chemically different monomeric units for the side branches and the backbone, etc. "Dendrimers" [17] ("starburst" or "combburst") are regular polymers that grow through multifunctional polymerization reactions starting from an initiator core and branching outward with a multiplication of the number of monomeric blocks from one generation to the next. Structure factors for "starburst" dendrimers are presented here as a guide to "build up" such quantities for other chain architectures. Polymer networks have been the subject of a number of SANS investigations [18]; however, as yet, their structure factor has not been successfully calculated making them inappropriate to include in this overview.

The paper is divided into the two main essential parts needed to work out a model for a generic polymer system: first, the single-chain structure factor, and then the inter-chain correlations contribution (also called "concentration" effects). Gaussian coil and random walk statistics are assumed for most of the paper, although, when possible, chain rigidity and chain swelling (or collapse) are included in order to show how these could be taken into account. A simple model, referred to as the "sliding rod" model [19], is used, for instance, to describe polymer chain stiffness between the gaussian coil limit and the rigid rod limit. Many other semiflexible chain models exist [20] but are not included here because of our focus on the simplest models and concepts possible. Monodisperse polymers are assumed throughout the paper. After discussing the two main pieces needed to work out a model (single-chain and inter-chain contributions), some specific examples are discussed and results taken from my recent research topics are included. Subject matters are covered at a basic tractable level.

2 Form Factors, Structure Factors

2.1 Definitions

In order to introduce some notation and definitions, we consider a polymer block with n monomers of segment length b. Defining $\mathbf{r}_{ij}$ as the interdistance between two monomers i and j in that block, and Q as the scattering wavenumber (often called scattering "vector"), various types of correlations can be considered depending on the number of summations involved:

$$E(Q) = \langle \exp[-i\mathbf{Q} \cdot \mathbf{r}_{1n}] \rangle \tag{2.1}$$

$$F(Q) = (1/n) \sum_{i=1}^{n} \langle \exp[-i\mathbf{Q} \cdot \mathbf{r}_{1i}] \rangle \tag{2.2}$$

$$P(Q) = (1/n^2) \sum_{i,j=1}^{n} \langle \exp[-i\mathbf{Q} \cdot \mathbf{r}_{ij}] \rangle. \tag{2.3}$$

$E(Q)$, $F(Q)$ and $P(Q)$ are the correlations between chain extremities, the form factor and the structure factor respectively and $\langle \cdots \rangle$ represents an average over all possible chain conformations. Note that these expressions are normalized such that they become unity at the zero Q limit.

The radius of gyration is defined as:

$$R_g^2 = (1/2n^2) \sum_{i,j=1}^{n} \langle r_{ij}^2 \rangle \tag{2.4}$$

and the low Q expansion of the structure factor is:

$$P(Q) = 1 - (QR_g)^2/3 \tag{2.5}$$

regardless of what model is used to describe chain statistics. However, the full chain correlations at finite Q can be calculated only after a chain model is specified. A few simple cases are considered here: flexible gaussian chains, rigid rods, "sliding rods" for semiflexible chains and freely-jointed chains.

2.2 Ideal Gaussian Polymer Blocks

Gaussian coils are characterized by a gaussian probability distribution [2] for the monomers and describe adequately flexible polymer blocks. Ideal chains follow random walk statistics, i.e.,

$$\langle r_{ij}^2 \rangle = b^2 |i - j| \tag{2.6}$$

(b being the segment length) as is the case for polymer melts, relaxed polymer solids and polymer solutions under theta conditions (when monomer–monomer and monomer–solvent interactions are equivalent). The polymer chain quickly "forgets" (after one step) about where its other segments are and the radius of gyration is $R_g = b(n/6)^{1/2}$.

In this case, a natural scattering variable is introduced as $\alpha = Q^2 b^2/6$ so that:

$$E_G(\alpha n) = \exp[-\alpha(n-1)] \qquad (2.7)$$

$$F_G(\alpha n) = (1/n)\sum_{i=1}^{n} \exp[-\alpha(i-1)]$$

$$= [1 - \exp(-\alpha n)]/[1 - \exp(-\alpha)]n \qquad (2.8)$$

$$P_G(\alpha n) = (1/n^2)\sum_{i,j=1}^{n} \exp[-\alpha|i-j|] \qquad (2.9)$$

$$= 1/n + 2\{1 - [1 - \exp(-\alpha n)]/[n(1 - \exp(-\alpha))]\}/$$
$$[n(1 - \exp(-\alpha))].$$

Note that for $\alpha \ll 1$ and $n \gg 1$, but keeping αn finite, more familiar form and structure factors are recovered:

$$F_G(\alpha n) = [1 - \exp(-\alpha n)]/\alpha n \qquad (2.10)$$

$$P_G(\alpha n) = 2[\exp(-\alpha n) - 1 + \alpha n]/(\alpha n)^2. \qquad (2.11)$$

This form for $P_G(\alpha n)$ is the widely used Debye function. Note that for homopolymer chains, n is large (large degree of polymerization) and α is small (SANS instruments do not "see" monomer chemistry) so that these last expressions can be used. However, for short block copolymers, n is not necessarily large and the more general equations are more appropriate to use.

2.3 Swollen/Collapsed Gaussian Polymer Blocks

Within Flory's mean field approximation, biased random walk statistics are characterized by:

$$\langle r_{ij}^2 \rangle = b^2|i-j|^{2\nu} \qquad (2.12)$$

where ν is the excluded volume parameter that takes on values between $\nu = 1/3$ for collapsed chains, to $\nu = 3/5$ for fully swollen chains, via $\nu = 1/2$ for ideal coils (note that fully swollen chains become ideal coils in 4 dimensional space). The radius of gyration is in this case: $R_g = b[n^{2\nu}/(2\nu + 1)(2\nu + 2)]^{1/2}$. The various correlation factors become:

$$E_G(\alpha n) = \exp[-\alpha(n-1)^{2\nu}] \qquad (2.13)$$

$$F_G(\alpha n) = (1/n)\sum_{i=1}^{n} \exp[-\alpha(i-1)^{2\nu}] \qquad (2.14)$$

$$P_G(\alpha n) = (1/n^2)\sum_{i,j=1}^{n} \exp[-\alpha|i-j|^{2\nu}]. \qquad (2.15)$$

Due to the fact that these progressions cannot be summed, the continuous chain limit ($\alpha \ll 1$ and $n \gg 1$, but keeping αn finite) is taken so that the following results can be obtained:

$$F_G(\alpha n) = [1/2\nu X^{1/2\nu}]\gamma(1/2\nu, X) \tag{2.16}$$

$$P_G(\alpha n) = [1/\nu X^{1/2\nu}]\{\gamma(1/2\nu, X) - [1/X^{1/2\nu}]\gamma(1/\nu, X)\} \tag{2.17}$$

where $X = \alpha n^{2\nu}$ has been defined and $\gamma(a, X)$ is the incomplete gamma function:

$$\gamma(a, X) = \int_0^X dt \exp(-t) t^{a-1}. \tag{2.18}$$

Note that $\gamma(a, \infty) = \Gamma(a)$ is the gamma function. This structure factor ($P_G(\alpha n)$) reproduces the Debye function in the limit $\nu = 1/2$. The high Q limit ($\alpha \gg 1$) of the structure factor is:

$$P_G(\alpha n) = [1/\nu X^{1/2\nu}]\{\Gamma(1/2\nu) - [1/X^{1/2\nu}]\Gamma(1/\nu)\} \tag{2.19}$$

which varies from $3\Gamma(3/2)/(\alpha^{3/2}n)$ for fully collapsed chains to $2/(\alpha n)$ for ideal chains to $5\Gamma(5/6)/3(\alpha^{5/6}n)$ for fully swollen chains. Flory introduced a characteristic transition temperature (called theta temperature and defined for infinite molecular weights) for which the monomer–monomer, solvent–solvent and monomer–solvent interactions are equivalent making the chain appear "ideal" (as if it were in a bulk environment). Below this temperature, chains start collapsing while above it, they start swelling.

2.4 The Temperature Blob Model

In cases of intermediate degrees of chain swelling, the temperature blob approach [21] has been useful in modeling excluded volume effects. It consists in defining a blob comprising a number n_τ of monomers that obey ideal chain statistics while the blobs themselves show excluded volume effects:

$$\begin{aligned} \langle r_{ij}^2 \rangle &= b^2|i - j|; & |i - j| < n_\tau \\ \langle r_{ij}^2 \rangle &= n_\tau b^2(|i - j|/n_\tau)^{2\nu}; & |i - j| > n_\tau. \end{aligned} \tag{2.20}$$

This approach involves two parameters (ν and n_τ) to describe chain swelling and is characterized by an unphysical break of chain statistics at $|i - j| = n_\tau$. Due to the awkwardness of the numerical generation of the structure factor when $\nu \neq 1/2$, a simple Debye function with swollen radius of gyration is often used (in an adhoc fashion) to fit scattering data from polymer solutions.

2.5 Rigid Rods

For rigid rods, the natural scattering variable is $\alpha = Qb$, and the conformational averaging becomes an averaging over orientations. Here also, in order to obtain compact analytical results, one has to go to the continuous chain limit ($\alpha \ll 1$

and n $\gg$ 1, but keeping αn finite) which gives:

$$E_R(\alpha n) = \sin(\alpha n)/\alpha n \tag{2.21}$$

$$F_R(\alpha n) = (1/n)\sum_i \sin(\alpha i)/\alpha i = Si(\alpha n)/\alpha n \tag{2.22}$$

$$P_R(\alpha n) = (1/n^2)\sum_{i,j}^{n} \sin[\alpha|i-j|]/[\alpha|i-j|]$$

$$= 2\{[\cos(\alpha n)-1]/\alpha n + Si(\alpha n)\}/\alpha n \tag{2.23}$$

where the sine integral function has been defined as:

$$Si(\alpha n) = \int_0^{\alpha n} dt \, \sin(t)/t. \tag{2.24}$$

Note that in cases where the large n assumption is not valid, precise form and structure factors should be generated by performing the summations numerically.

Various models are available [20] to describe semiflexible chains. Some are based on expansions either close to the gaussian coil or to the rigid rod limits, while others interpolate between these two chain stiffness limits. One of these, the sliding rod model [19], is described here because of its inherent simplicity.

2.6 The Sliding Rod Model

The sliding rod model assumes that the chain behaves as a rigid rod for contour lengths corresponding to a characteristic number n_c of monomers (bn_c is used as a stiffness parameter reminiscent of the Kuhn length) whereas longer chain portions follow ideal chain gaussian statistics (flexible coils). The structure factor is given by:

$$P_{SR}(\alpha_1 n, \alpha_2 n) = (1/n^2)\left\{ \sum_{i-j=1}^{n_c} \sin[\alpha_1|i-j|]/\alpha_1|i-j| \right.$$

$$\left. + \sum_{i-j=n_c+1}^{n} \exp(-\alpha_2|i-j|)\right\} \tag{2.25}$$

where $\alpha_1 = Qb$ and $\alpha_2 = Q^2 b^2/6$. Defining structure factors for a rigid rod $P_R(\alpha_1 n)$ and a gaussian coil $P_G(\alpha_2 n)$ (see above), one obtains:

$$P_{SR}(\alpha_1 n, \alpha_2 n) = (1/n^2)\{n_c^2 P_R(\alpha_1 n)$$

$$+ (n - n_c - 1)^2 P_G(\alpha_2 n - \alpha_2 n_c - \alpha_2)\}. \tag{2.26}$$

2.7 Freely-Jointed Chains

Freely jointed chains are another form of flexible chains because of the free rotation of the universal joint between monomers; however, monomers are assumed to be rigid "sticks" of size b. In this case, the structure factor can be calculated:

$$P_{FJC}(\alpha n) = \{[1 + j_0(\alpha)]n/[1 - j_0(\alpha)] - 2j_0(\alpha)[1 - j_0^n(\alpha)]/ $$
$$[1 - j_0(\alpha)]^2\}/n^2 \qquad (2.27)$$

where $\alpha = Qb$ and $j_0(\alpha) = \sin(\alpha)/\alpha$ is the spherical Bessel function of order zero.

Now that self correlations within the same block (or chain) have been calculated, we will introduce cross correlations between different blocks.

2.8 Inter-Block Cross Correlations

Consider two polymer blocks A and C with n_A and n_C monomers of segment lengths b_A and b_C in each, separated by a third block B with n_B monomers of segment length b_B. Correlations between the A and C blocks are [22]:

$$P(\alpha_A n_A, \alpha_C n_C) = (1/n_A n_C) \sum_{i=1}^{n_A} \sum_{j=1}^{n_C} \langle \exp[-\mathbf{Q} \cdot \mathbf{r}_{ij}] \rangle$$
$$= F(\alpha_A n_A) E(\alpha_B n_B) F(\alpha_C n_C) \qquad (2.28)$$

where completely free joints (random bond angles) have been assumed between blocks A/B and B/C and angular correlations have been neglected. $E(\alpha n)$ and $F(\alpha n)$ could be for rigid rods or gaussian coils depending on the stiffness of the A, B, and C blocks.

Now that the needed tools for calculating intra-block and interblock correlations have been developed, we will consider various polymer architectures in the next sections.

3 Structure Factors for Various Chain Architectures

3.1 Linear Chain

Consider a linear chain with N monomers of segment length b. The static scattering function is defined as: $S(Q) = N^2 P(\alpha N)$. For convenience in notation, it can also be written as:

$$S(Q) = N + N^2 Q[\alpha, N] \qquad (3.1)$$

where:

$$Q[\alpha, N] = (2/N^2) \sum_{k=1}^{N} (N - k) E(\alpha k) \qquad (3.2)$$

represents the non-self $(i \neq j)$ correlations. Here, $\alpha = Q^2 b^2/6$ if the chain is flexible or $\alpha = Qb$ if it is a rigid rod. Note that:

$$Q[\alpha, N] = P(\alpha, N) - 1/N \tag{3.3}$$

and that $Q[0, N] = (N - 1)/N$. Efforts are being made to keep N finite (for the sake of describing short blocks) to the extent possible.

3.2 Ring Homolymer

Since modeling rigid rings is trivial. We assume gaussian statistics (ideal chains). Consider a ring polymer of N monomers of segment length b. The static scattering function is given by the following identical expressions:

$$S(Q) = \sum_{i, j}^{N} \exp[-\alpha|i - j|(1 - |i - j|/N)] \tag{3.4a}$$

$$= N + 2N \sum_{k}^{N} (1 - k/N)\exp[-\alpha k(1 - k/N)] \tag{3.4b}$$

$$= N \sum_{k}^{N} \exp[-\alpha k(1 - k/N)] \tag{3.4c}$$

S(Q) can be readily calculated [23] in the continuous chain limit ($\alpha \ll 1$ and $N \gg 1$, but keeping αN finite) as:

$$N^2[\exp(-U^2)/U] \int_{0}^{U} dt \exp(t^2) \tag{3.5}$$

where $U = (\alpha N)^{1/2}/2$ and the Dawson integral can be generated numerically (it is part of some computer libraries).

3.3 Diblock Linear Copolymer

Consider a diblock A and B linear copolymer (n_A, n_B, α_A, α_B). The static scattering function S(Q) is the sum of three contributions:

$$S(Q) = S_{AA}(Q) + S_{BB}(Q) + 2S_{AB}(Q) \tag{3.6}$$

where:

$$S_{AA}(Q) = n_A^2 P(\alpha_A n_A) \tag{3.7a}$$

$$S_{BB}(Q) = n_B^2 P(\alpha_B n_B) \tag{3.7b}$$

$$S_{AB}(Q) = n_A n_B F(\alpha_A n_A) F(\alpha_B n_B). \tag{3.7c}$$

Note that $S_{AB}(Q)$ can also be written as:

$$S_{AB}(Q) = [n^2 P(\alpha n) - n_A^2 P_A(\alpha_A n_A) - n_B^2 P_B(\alpha_B n_B)]/2 \tag{3.7d}$$

where $\alpha = (\alpha_A n_A + \alpha_B n_B)/n$ and $n = n_A + n_B$.

3.4 Diblock Ring Copolymer

Consider a diblock A and B cyclic copolymer (n_A, n_B, α_A, α_B). As in the linear case, the scattering function is the sum of three contributions with, however:

$$S_{AA}(Q) = \sum_{i,j}^{n_A} \exp[-\alpha_A |i-j|(1-|i-j|/n)] \tag{3.8a}$$

$$= n_A + 2n_A \sum_k^{n_A} (1-k/n_A)\exp[-\alpha_A k(1-k/n)] \tag{3.8b}$$

which becomes in the continuous limit:

$$= 2n_A^2 \exp(-\alpha_A n/4) \int_0^1 ds(1-s)\exp[\alpha_A n(n_A s/n - 1/2)^2] \tag{3.8c}$$

where $n = n_A + n_B$. As for homopolymer rings, this integral has to be performed numerically.

The cross correlations structure factor $S_{AB}(Q)$ can be obtained using Eq. (3.7d) where $n^2 P(\alpha n)$ becomes the structure factor for the full n monomer ring.

3.5 Alternating Copolymer

Consider a regularly alternating linear copolymer (Fig. 1) of N_A blocks A and N_B blocks B with N being the total number of blocks ($N = N_A + N_B$). Each A(B) block is comprised of n_A (n_B) monomers. Two cases will be considered: (1) N is odd and (2) N is even.

When N is odd, $N_A = (N+1)/2$ and $N_B = (N-1)/2$; the A/A scattering functions are:

$$S_{AA}(Q) = N_A S_{AA}^S(Q) + N_A^2 S_{AA}^I(Q). \tag{3.9}$$

The self block correlations are straightforward to evaluate:

$$S_{AA}^S(Q) = n_A^2 P(\alpha_A n_A) \tag{3.10}$$

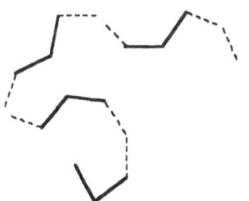

Fig. 1. Regularly alternating block copolymer

and the inter block correlations are:

$$S^I_{AA}(Q) = 2(n^2_A/N^2_A) \sum_{k=1,2}^{N_A} (N_A - k)F^2(\alpha_A n_A)E(\alpha_A n_A)^{(k-1)}E(\alpha_B n_B)^k$$

(3.11)

where $(N_A - k)$ is the number of A/A block combinations that are separated by 2k blocks. In the case of gaussian chains:

$$S^I_{AA}(Q) = n^2_A F^2_G(\alpha_A n_A)E_G(-\alpha_A n_A)Q_G[\alpha_A n_A + \alpha_B n_B, N_A].$$ (3.12)

Similarly for the A/B scattering function:

$$S_{AB}(Q) = 2(n_A n_B/N_A N_B) \sum_{k=1,2}^{N_A} (N_A - k)F(\alpha_A n_A)F(\alpha_B n_B)$$

$$E(\alpha_A n_A)^{(k-1)}E(\alpha_B n_B)^{(k-1)}$$

(3.13)

and for gaussian chains:

$$S_{AB}(Q) = (N_A N_B n_A n_B)F_G(\alpha_A n_A)F_G(\alpha_B n_B)$$

$$E_G(-\alpha_A n_A)E_G(-\alpha_B n_B)Q_G[\alpha_A n_A + \alpha_B n_B, N_A].$$ (3.14)

When N is even, $N_A = N_B = N/2$ and the corresponding results are:

$$S^I_{AA}(Q) = n^2_A F^2_G(\alpha_A n_A)E_G(-\alpha_A n_A)Q_G[\alpha_A n_A + \alpha_B n_B, N_A]$$ (3.15)

$$S_{AB}(Q) = (N_A N_B n_A n_B)F_G(\alpha_A n_A)F_G(\alpha_B n_B)E_G(-\alpha_A n_A)E_G(-\alpha_B n_B)$$

$$\{Q_G[\alpha_A n_A + \alpha_B n_B, N_A + 1]$$

$$+ Q_G[\alpha_A n_A + \alpha_B n_B, N_A]\}/2.$$ (3.16)

The B/B and B/A scattering functions are obtained by interchanging the A and B indices. The total scattering function is the sum of all contributions:

$$S(Q) = S_{AA}(Q) + S_{BB}(Q) + 2S_{AB}(Q)$$ (3.17)

provided that the monomers have the same scattering length densities. If not, then the scattering intensity is obtained by weighing the static scattering functions by the appropriate contrast factors (as will be discussed later).

3.6 Star Branched Copolymer

Consider a star branched polymer with N_b branches (Fig. 2). Each branch is composed of two monomeric blocks A and B (copolymer) of n_A and n_B monomers respectively (the A block is assumed to be at the outside of the branch). The single-branch correlations $S_{sb}(Q)$ involve the scattering functions $S^{sb}_{AA}(Q)$, $S^{sb}_{BB}(Q)$, and $S^{sb}_{AB}(Q)$ that were derived [24] for diblock copolymers:

$$S_{sb}(Q) = S^{sb}_{AA}(Q) + S^{sb}_{BB}(Q) + 2S^{sb}_{AB}(Q)$$ (3.18)

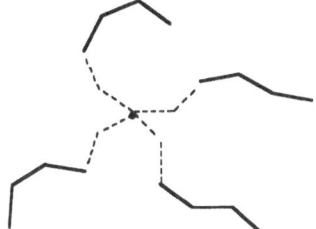

Fig. 2. Four-arm star branched copolymer with three outer A blocks and two inner B blocks

$$S_{AA}^{sb}(Q) = n_A^2 F^2(\alpha_A n_A)$$

$$S_{BB}^{sb}(Q) = n_B^2 F^2(\alpha_B n_B)$$ (3.19)

$$S_{AB}^{sb}(Q) = n_A n_B F(\alpha_A n_A) F(\alpha_B n_B),$$

while the inter-branch correlations:

$$S_{ib}(Q) = S_{AA}^{ib}(Q) + S_{BB}^{ib}(Q) + 2S_{AB}^{ib}(Q)$$ (3.20)

involve triblock correlations that were calculated for alternating copolymers ($N = 3$ in the preceding section) with n_A monomers in each of the two outside A blocks and $2n_B$ monomers in block B:

$$S_{AA}^{ib}(Q) = n_A^2 F^2(\alpha_A n_A) E(\alpha_B 2n_B)$$

$$S_{BB}^{ib}(Q) = n_B^2 F^2(\alpha_B n_B)$$ (3.21)

$$S_{AB}^{ib}(Q) = n_A n_B F(\alpha_A n_A) E(\alpha_B n_B) F(\alpha_B n_B) .$$

The total static scattering function is therefore the sum of the two contributions:

$$S(Q) = N_b S_{sb}(Q) + N_b(N_b - 1)S_{ib}(Q).$$ (3.22)

Here also, the various self and cross correlations have to be weighed by the contrast factors in order to obtain the scattered intensity.

3.7 Copolymer Comb

Consider a regular copolymer "comb" (Fig. 3) made of N_A side branches (n_A monomers in each) and N_B backbone blocks (n_B monomers in each). Note that usually $N_B = N_A + 1$ and that the backbone units are called blocks only for convenience. The various correlation terms involved are [22]:

$$S_{AA}(Q) = N_A S_{AA}^s(Q) + N_A(N_A - 1)S_{AA}^I(Q)$$ (3.23)

$$S_{AA}^s(Q) = n_A^2 P(\alpha_A n_A)$$ (3.24)

$$S_{AA}^I(Q) = n_A^2 F^2(\alpha_A n_A) Q[\alpha_B n_B, N_A]$$

$$S_{BB}(Q) = (N_B n_B)^2 P(\alpha_B N_B n_B)$$ (3.25)

Fig. 3. Copolymer comb with four side branches equally spaced along the chain backbone

$$S_{AB}(Q) = 2(n_A n_B/N_A N_B) F(\alpha_A n_A) \sum_{k}^{N_B - 1} k F(\alpha_B n_B k) . \tag{3.26}$$

If the B monomers are flexible (gaussian statistics) $S_{AB}(Q)$ becomes:

$$S_{AB}(Q) = 2(n_A n_B/N_A N_B) F(\alpha_A n_A) N_A \{1 - F_G(\alpha_B n_B N_A)\}/$$
$$[\exp(\alpha_B n_B) - 1] \tag{3.27}$$

or if they are rigid, then:

$$S_{AB}(Q) = 2(n_A n_B/N_A N_B) F(\alpha_A n_A)(N_A/\alpha_B n_B)$$
$$\times \{[1 - \cos(\alpha_B n_B N_A)] \sin(\alpha_B n_B)$$
$$+ [1 - \cos(\alpha_B n_B)] \sin(\alpha_B n_B N_A)\}/\{[1 - \cos(\alpha_B n_B)]^2$$
$$+ \sin^2(\alpha_B n_B)\} \tag{3.28}$$

regardless of whether the A monomers are flexible (use $F_G(\alpha_A n_A)$) or rigid (use $F_R(\alpha_A n_A)$). Note that orientational correlations have been neglected.

3.8 Starburst Dendrimer

Consider a regular starburst dendrimer [17] formed of N_b branches. Each branch is formed of N generations of monomeric blocks going from the first generation at the core to higher generations outside. The number of blocks is multiplied by a factor f (usually f = 2) in going from one generation to the next. Note that the "functionality" parameter is defined here as f + 1. Each block is composed of n monomers forming gaussian links with segment length b.

The calculations of the various structure factors for a dendrimer are rather straightforward [25], but somewhat tedious. There are four main contributions to these correlations: (1) one intrabranch self-correlations part, S_{sb}^s, (2) one intra-branch cross-correlations part between blocks that originate from the same stem, S_{sb}^f, (3) one intra-branch cross-correlations part between blocks that originate from different stems, S_{sb}^a, and (4) one interbranch correlations part S_{ib}. These various correlations are sketched in Fig. 4.

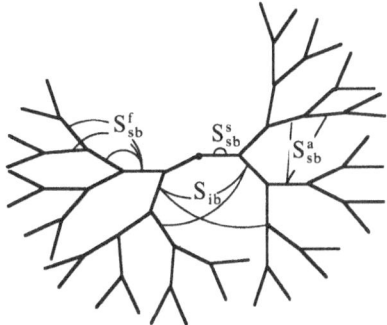

Fig. 4. Dendrimer polymer gel with two branches, five generations and a doubling of the number of units at each generation

The intra-branch self correlations term in the scattering function involves monomer–monomer correlations within the same block. Since there are $f^{(k-1)}$ blocks in generation k, the total number of blocks per branch, n_B, is:

$$n_B = \sum_{k=1}^{N} f^{k-1} = (f^N - 1)/(f - 1) \tag{3.29}$$

therefore giving (gaussian monomers are assumed):

$$S_{sb}^s(Q) = n^2 P_G(\alpha n) . \tag{3.30}$$

The intra-branch cross-correlations between blocks that originate at the same stem involve summations over blocks in generations k and l respectively and form factors internal to each block:

$$S_{sb}^f(Q) = 2n^2 [F_G(\alpha n)]^2 \sum_{k=1}^{N} f^{k-1} \sum_{r=k+1}^{N} f^{r-k} \exp[-\alpha n(r - k - 1) . \tag{3.31}$$

These summations can be easily performed giving an analytical expression:

$$S_{sb}^f(Q) = 2n^2 \{[F_G(\alpha n)]^2/[f\exp(-\alpha n) - 1]\}$$
$$\times \{f^N \exp(-\alpha nN)[\exp(\alpha nN) - \exp(\alpha n)]/$$
$$[\exp(\alpha n) - 1] - f(f^N - 1)/(f - 1)\}. \tag{3.32}$$

Similarly, for the intra-branch cross-correlations between blocks that originate from different stems, three summations are involved: the previous two (over k and r) and a third summation over the number of stem points (m) that have to be crossed in order to join the two blocks under consideration:

$$S_{sb}^a(Q) = 2n^2 [F_G(\alpha n)]^2 \sum_{k=2}^{N} f^{k-1} \sum_{m=1,3}^{2k-3} (f - 1) f^{(m-1)/2}$$
$$\times \exp[-\alpha n(m - 1)]$$
$$\times \left\{1 + 2 \sum_{r=k+1}^{N} f^{r-k} \exp[-\alpha n(r - k)]\right\}. \tag{3.33}$$

In other words, in going from block k to a block r, one has to meet m stems with $(f - 1)f^{(m-1)/2}$ as the number of possibilities. The summations, here also, can be performed giving:

$$S_{sb}^a(Q) = 2n^2[F_G(\alpha n)]^2\{(f-1)/[f\exp(-2\alpha n) - 1]\}$$
$$\times \{A(Q) + B(Q)\} \qquad (3.34)$$

$$A(Q) = [f^{2N}\exp(-2\alpha nN) - f^2\exp(-2\alpha n)]/$$
$$[f^2\exp(-2\alpha n) - 1] - (f^N - f)/(f - 1)$$

$$B(Q) = 2\{f^N\exp(-\alpha nN)[f^N\exp(-\alpha nN) - f\exp(-\alpha n)]/$$
$$[f\exp(-\alpha n) - 1]$$
$$- f\exp(-\alpha n)[f^{2N}\exp(-2\alpha nN) - f^2\exp(-2\alpha n)]/$$
$$[f^2\exp(-2\alpha n) - 1] - f^N\exp(-\alpha nN)[\exp(\alpha nN) - \exp(\alpha n)]/$$
$$[\exp(\alpha n) - 1] + f\exp(-\alpha n)(f^N - f)/$$
$$(f - 1)\}/[f\exp(-\alpha n) - 1] .$$

Note that this term is proportional to $(f - 1)$ and goes to zero for star-branched polymers $(f = 1)$.

Finally, the inter-branch correlations are:

$$S_{ib}(Q) = n^2[F_G(\alpha n)]^2 \sum_{k=1}^{N} f^{k-1} \sum_{r=1}^{N} f^{r-1} \exp[-\alpha n(r + k - 2)] \quad (3.35)$$

and are summed up to give:

$$S_{ib}(Q) = n^2[F_G(\alpha n)]^2[f^N\exp(-\alpha nN) - 1]^2/$$
$$(f\exp(-\alpha n) - 1)^2. \qquad (3.36)$$

The total scattering function is the sum of all of these partial structure factors:

$$S(Q) = N_b[S_{sb}^s(Q) + S_{sb}^f(Q) + S_{sb}^a(Q)] + N_b(N_b - 1)S_{ib}(Q) . \quad (3.37)$$

This scattering function goes to the square of the total number of monomers in the gel, $(nN_bn_B)^2$, at the zero Q limit as it should.

The calculations presented here agree with those of Burchard et al. [26] who used cascade theory to investigate the case corresponding to $f = 2$. In order to derive these results, we have assumed ideal gaussian monomer blocks that can cross each other ("phantom" chains). This is unrealistic for high functionality where the monomer density becomes so high that the reaction sites get screened therefore stopping the polymerization reaction after about 10 generations. Moreover, monomer blocks that are far from the core find themselves in "stretched" configurations (with enhanced excluded volume) due to the lack of available space. Excluded volume can be incorporated either at the outset in which case the summations would have to be performed numerically or in an ad-hoc fashion (as mentioned before). The approach presented here is, however,

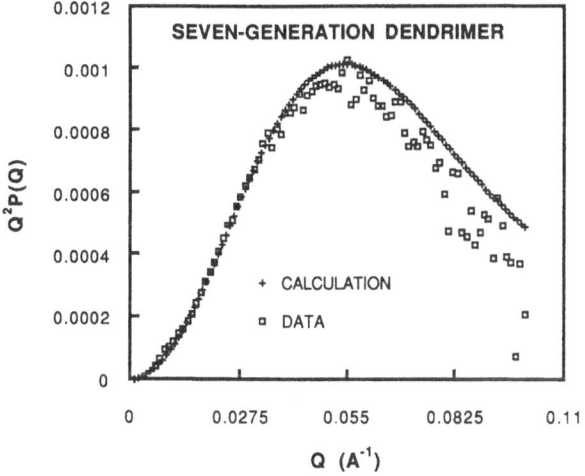

Fig. 5. Kratky plot for SANS data taken from solutions of starburst polyamidoamine dendrimers with seven generations in deuterated water (extrapolated to zero concentration) and comparison with the gaussian blocks dendrimer model with b = 0.48 nm, n = 10, N = 7, N_b = 3 and f = 2. Salt was added to the solution in order to screen out the coulomb interactions

a first start in the modeling of such complicated structures as shown in Fig. 5 where SANS data taken from solutions of starburst polyamidoamine dendrimers with seven generations in deuterated water are compared with various models including the gaussian blocks dendrimer model described here (with b = 0.48 nm, n = 10, N = 7, N_b = 3 and f = 2). In order to reach the single dendrimer scattering limit, low concentrations (less than 1% volume fraction) were measured and data were extrapolated to zero concentration.

4 Dilute Polymer Solutions

The single-chain structure factors calculated in the previous sections correspond to the infinite dilution limit. This limit also corresponds to zero scattering intensity and is not useful so that concentration effects have to be included in the modeling of polymer solutions. First, Zimm's single-contact approximation [5] is reviewed for dilute polymer solutions; then, a slight extension of that formula which applies to semidilute solutions, is discussed.

4.1 Zimm's Single Contact Approximation

Consider a dilute polymer solution with N_p polymer molecules of N monomers each. The polymer volume fraction ϕ_p is often used as the concentration parameter; it is given by: $\phi_p = NN_pv_p/(NN_pv_p + N_sv_s)$, where v_p and v_s are the monomeric and solvent molecule volumes respectively and N_s is the number of solvent molecules. Note that, sometime, the polymer concentration $C_p = N_p/V$ ($V = NN_p v_p + N_s v_s$ being the total sample volume) is used instead.

Zimm's approximation [5] assumes that inter-chain interactions occur only through single contacts. Given two monomers i and j that belong to two different chains (say, called 1 and 2), the two-chain distribution function is:

$$P(\mathbf{r}_{1i}, \mathbf{r}_{2j}) = P(\mathbf{r}_{1i}) P(\mathbf{r}_{2j}) \left\{ 1 - v \sum_{k,1}^{N} \delta(\mathbf{r}_{1k,21}) \right\} \tag{4.1}$$

where $P(\mathbf{r})$ is the single-chain distribution function, $\mathbf{r}_{1k,21} = \mathbf{r}_{1k} - \mathbf{r}_{21}$, v is the excluded volume during the binary interaction, and $\delta(\mathbf{r})$ is the Dirac delta function. The scattering function for the whole sample is the sum of two (single-chain and inter-chain) contributions which are:

$$S(Q) = N_p S_{sc}(Q) + N_p(N_p - 1) S_{ic}(Q) \tag{4.2}$$

$$S_{sc}(Q) = N^2 P(\alpha N) \tag{4.3}$$

$$S_{ic}(Q) = -(v/V) N^4 P^2(\alpha N)$$

where we have neglected the Fourier transform of the average polymer density (which is identically zero except at $Q = 0$ which is an experimentally irrelevant limit):

$$\langle \rho(Q) \rangle = (1/N) \sum_{i}^{N} \langle \exp(-i\mathbf{Q} \cdot \mathbf{r}_i) \rangle \sim \delta(\mathbf{Q}). \tag{4.4}$$

This is equivalent to assuming a constant polymer density ($\langle \rho(\mathbf{r}) \rangle$) in configuration space. The second virial coefficient A_2 is related to v as follows:

$$N^2 v = 2M_w^2 A_2/N_{av} \tag{4.5}$$

where M_w and N_{av} are the polymer molecular weight and Avogadro's number respectively. The scattering function is therefore (N_p being always large):

$$S(Q) = N_p N^2 P(\alpha N) \{ 1 - vC_p N^2 P(\alpha N) \}. \tag{4.6}$$

This formula describes dilute solutions ($C_p < 1/vN^2$) fairly well despite the fact that it ignores macromolecular shape changes during binary chain interactions.

4.2 The Inverse Zimm Formula

The inverse Zimm formula involves the following approximate form:

$$S^{-1}(Q) = \{ 1 + vC_p N^2 P(\alpha N) \}/N_p N^2 P(\alpha N) .$$
$$= 1/N_p N^2 P(\alpha N) + v/V. \tag{4.7a}$$

or in terms of volume fractions:

$$S^{-1}(Q) = (v_p^2/V) \{ 1/N\phi_p v_p P(\alpha N) + v/v_p^2 \}. \tag{4.7b}$$

This formula is the basis of the Zimm Plot which consists in plotting the inverse of the scattering intensity, $S^{-1}(Q)$, vs Q^2 which shows a linear variation at low Q and in dilute solutions. Extrapolated values ($Q \rightarrow 0$, $C_p \rightarrow 0$) of the intercept and the slope yield the degree of polymerization N and the excluded volume v (or second virial coefficient A_2) respectively. Zimm's formula can describe scattering data accurately well into the semidilute concentration region. This region is defined for concentrations above an overlap concentration C_p^* which is defined in either of the two following ways:

$$C_p^* = M_w/N_{av}R_g^3 \quad \text{or} \quad C_p^* = 3M_w/4\pi N_{av}R_g^3 \tag{4.8}$$

where R_g is the macromolecular radius of gyration defined before. Note that this overlap concentration does not correspond to a "critical" state of the system. The inverse Zimm formula, which is the "approximation of an approximation", works surprisingly well due to the fact that the second approximation, $(1 - x) \sim 1/(1 + x)$, re-sums higher order terms which account for series of single-contacts. It applies whether gaussian or rigid rod statistics are assumed. Actually, for rigid rods, Zimm's single contact approximation works better because the third and higher virial coefficients are small. Zimm's single contact approximation, however, neglects orientational correlations which can be important in semidilute solutions of rigid rods. Because it does not account for all possible correlations, the inverse Zimm scattering formula is essentially a mean field treatment.

Some misgivings as to the use of such a mean field approach to describe polymer solutions have been presented in the literature [4, 9]. For instance, the mean field approach cannot explain the experimentally observed molecular weight dependence of the excluded volume v and does not account for entanglements. A renormalization group description of polymer solutions exists in the literature [10]. It is more successful than the mean field description for dilute and semidilute solutions where concentration fluctuations are non-negligible (especially in good solvents). For instance, it can account for the molecular weight dependence of the excluded volume v. However, due to my lack of knowledge concerning this research area, these are not included here. Moreover, scaling theory arguments [4, 9] and Monte Carlo predictions [27] are available for dilute and semidilute polymer solutions. These, however, cannot be used to fit experimental data (scaling theory cannot predict prefactors and Monte Carlo calculations can be used only to compare trends).

5 Concentrated Polymer Solutions

A characteristic concentration C_p^{**} can be defined [28] to separate the semidilute and concentrated regions:

$$C_p^{**} = C_p^* (R^2/R_\theta^2)^4 \tag{5.1}$$

where R^2 and R_θ^2 are the end-to-end chain distances at zero concentration and under theta conditions respectively. As the concentration increases, excluded volume effects start being screened so that they become negligible in polymer melts whereby random walk statistics are recovered (zero second virial coefficient). In semidilute and concentrated regions, one could define concentration blobs between entanglement points. Chains would display excluded volume effects within a concentration blob but the blobs themselves would follow random walk statistics. This approach will not be pursued here. Instead, we will discuss a justification of Zimm's inverse formula in terms of series of single-contacts (the Benoit–Benmouna model [29]) and review the "high concentration method" [6–8] which is an alternative to the Zimm approach and permits the separation of single-chain and interchain structure factors by performing measurements from samples with different deuterated polymer fractions (keeping the total polymer concentration constant).

5.1 The Benoit–Benmouna Model

For non-dilute concentrations ($C_p > 1/vN^2$), a binary interaction between two macromolecules can occur either through a direct contact, or through a series of contacts [29] with other chains (third, fourth, etc.). If k chains are involved, the scattering function is:

$$S(Q) = N_p N^2 P(\alpha N) \{ 1 - v C_p N^2 P(\alpha N) + v^2 C_p^2 N^4 P^2(\alpha N)$$
$$+ \cdots + (-v)^k C_p^k N^{2k} P^k(\alpha N) \}. \tag{5.2}$$

This series can be re-summed to give the inverse Zimm formula:

$$S(Q) = N_p N^2 P(\alpha N) / \{ 1 + v C_p N^2 P(\alpha N) \} \tag{5.3}$$

RANDOM PHASE APPROXIMATION

Interactions Included

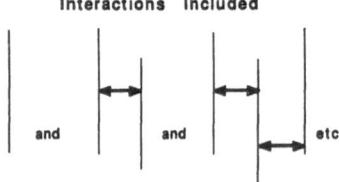

and   and   etc

Interactions Neglected

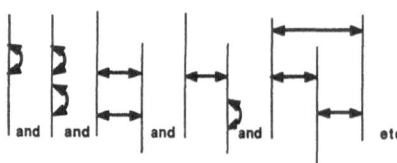

and   and   and   and   etc

Fig. 6. Schematic representation of the interchain interactions that are included or neglected in the random phase approximation

which still does not contain contributions from multiple contacts between two chains. Also neglected are "loop" (sometime called "ring") interactions within a chain, multiple contacts with other chains, etc. (see Fig. 6). The loop interactions are the ones that account for excluded volume effects and the multiple contacts properly account for the molecular weight dependence of the excluded volume (second virial coefficient) in the renormalization group theory [10]. Moreover, ternary interactions (neglected in the mean field approach) are non negligible in highly branched systems (such as star polymers with a large number of branches).

Effects due to macromolecular shape changes during single-contact interactions can, within the Benoit–Benmouna theory, be included in an ad-hoc fashion by renormalizing the single-chain structure factor to make it concentration dependent. This approach is often used to describe polymer solutions up to the concentrated region.

5.2 The High Concentration Method

Consider a polymer solution consisting of protonated and deuterated polymers (concentrations C_H and C_D respectively) that have the same degree of polymerization N. The scattered intensity is proportional to the structure factors for the polymer–polymer correlations $S_{HH}(Q)$, $S_{DD}(Q)$, $S_{HD}(Q)$, polymer–solvent correlations $S_{PS}(Q)$, and solvent–solvent correlations $S_{SS}(Q)$. The protonated, deuterated and solvent molecule scattering lengths are called a_H, a_D and a_S and $\{ap\}$ is the average polymer scattering length. The scattered intensity can be expressed in terms of an absolute cross section as:

$$d\Sigma(Q)/d\Omega = [a_D^2 S_{DD}(Q) + a_H^2 S_{HH}(Q) + 2a_D a_H S_{DH}(Q)$$
$$+ 2a_S\{a_p\} S_{PS}(Q) + a_S^2 S_{SS}(Q)]/V . \qquad (5.4a)$$

Assuming an incompressible polymer solution allows the elimination of the last two terms:

$$d\Sigma(Q)/d\Omega = [(a_D/v_D - a_S/v_S)^2 v_D^2 S_{DD}(Q) + (a_H/v_H - a_S/v_S)^2 v_H^2 S_{HH}(Q)$$
$$+ 2(a_D/v_D - a_S/v_S)(a_H/v_H - a_S/v_S) v_D v_H S_{DH}(Q)]/V. \qquad (5.4b)$$

The various structure factors can be split, here also, into single-chain ($P_D^S(Q)$, etc.) and interchain ($P_{DD}^I(Q)$, etc.) parts:

$$v_D^2 S_{DD}(Q)/V = N_D \phi_D v_D [P_D^S(Q) + \phi_D P_{DD}^I(Q)]$$
$$v_H^2 S_{HH}(Q)/V = N_H \phi_H v_H [P_H^S(Q) + \phi_H P_{HH}^I(Q)] \qquad (5.5)$$
$$v_D v_H S_{DH}(Q)/V = (N_D \phi_D v_D N_D \phi_D v_D)^{1/2} (\phi_D \phi_H)^{1/2} P_{DH}^I(Q) .$$

The high concentration method assumes that deuteration does not change chain conformations and interactions (provided that the total polymer concentration is kept constant); i.e., that $P_D^S(Q) = P_H^S(Q)$ (call it $P_S(Q)$) and

$P^I_{DD}(Q) = P^I_{HH}(Q) = P^I_{DH}(Q)$ (call it $P_I(Q)$) along with $N_D = N_H$ (call it N) and $v_D = v_H$ (call it v_p). Defining a "total" polymer–polymer structure factor $P_T(Q) = P_S(Q) + \phi_P P_I(Q)$, the main result is obtained in terms of the various contrast factors:

$$\{A^2_P\} = (a_D/v_P - a_S/v_S)^2 \, \phi_D/\phi_P + (a_H/v_P - a_S/v_S)^2 \, \phi_H/\phi_P \qquad (5.6)$$

$$\{A_P\}^2 = [(a_D/v_P - a_S/v_S)\phi_D/\phi_P + (a_H/v_P - a_S/v_S)\phi_H/\phi_P]^2$$

as:

$$d\Sigma(Q)/d\Omega = [\{A^2_P\} - \{A_P\}^2] N\phi_P v_P P_S(Q) + \{A_P\}^2 N\phi_P v_P P_T(Q) \, . \qquad (5.7)$$

where we have defined the following: $\{ \cdots \}$ is a composition averaging and ϕ_P is the total polymer volume fraction ($\phi_P = \phi_D + \phi_H$), so that $\{a_P\} = a_H \phi_H/\phi_P + a_D \phi_D/\phi_P$ is the average polymer scattering length.

The first contrast factor can be simplified as:

$$[\{A^2_P\} - \{A_P\}^2] = (a_H/v_H - a_D/v_D)^2 \, \phi_D \phi_H/\phi^2_P \qquad (5.8)$$

where ϕ's are the volume fractions. The structure factors $P_S(Q)$ and $P_T(Q)$ can, therefore, be obtained by performing two measurements (where only ϕ_D/ϕ_H is varied, keeping $\phi_P = \phi_D + \phi_H$ constant). Because $P_T(Q)$ represents correlations from all monomers with equal weighting, it has a weak Q dependence. Note that, if the incompressibility assumption had not been made, one would need four different sample compositions instead, in order to determine the four unknown structure factors $P_S(Q)$, $P_T(Q)$, $S_{PS}(Q)$ and $S_{SS}(Q)$.

This method applies for whatever polymer concentration. In practice, it is preferable to use high concentrations in order to increase the signal-to-noise ratio and therefore minimize counting time. However, it can also be applied to semidilute or even dilute solutions where Zimm plots are useful. It also applies not only to linear polymers but also to any form of chain architecture and to deuterated/protonated mixtures in non-solvent matrices such as polymer blends or polymer networks provided that changing the deuterated fraction does not change the homogeneous nature of the mixture (i.e., no change to the chain conformations and interactions).

6 Multicomponent Blends of Flexible Homopolymers and Copolymers

6.1 The de Gennes Formula

Consider a binary polymer blend (A and B components) of gaussian chains with degrees of polymerization N_A, N_B, volume fractions ϕ_A, ϕ_B, and monomeric volumes v_A, v_B, respectively. When the blend is a homogeneous phase mixture, de Gennes [3–4] used the Random Phase Approximation (RPA) to derive an expression for the structure factors of the fully interacting system, $S_{AA}(Q)$, $S_{BB}(Q)$, $S_{AB}(Q)$, in terms of those of the "ideal" (non-interacting) system $S^0_{AA}(Q)$, $S^0_{BB}(Q)$. $S^0_{AA}(Q)$ and $S^0_{BB}(Q)$ are the single chain structure factors (note that $S^0_{AB}(Q) = 0$ except for A/B copolymers). Assuming effective interaction potentials W_{AA}, W_{BB} and W_{AB} between A and B monomers, the linear responses of the fluctuating densities $\langle \rho_A(Q) \rangle$ and $\langle \rho_B(Q) \rangle$ to externally applied (weakly perturbing) potentials U_A and U_B are given by:

$$\langle \rho_A(Q) \rangle = - X^0_{AA}(Q)[U_A + W_{AA}\langle \rho_A(Q) \rangle + W_{AB}\langle \rho_B(Q) \rangle]/k_B T$$
$$\langle \rho_B(Q) \rangle = - X^0_{BB}(Q)[U_B + W_{BA}\langle \rho_A(Q) \rangle + W_{BB}\langle \rho_B(Q) \rangle]/k_B T \qquad (6.1)$$

where $X^0_{AA}(Q)$, etc., represent "bare" response functions. The isothermal incompressibility ($\langle \rho_A(Q) \rangle + \langle \rho_B(Q) \rangle = 0$) condition allows the elimination of one of the equations. The fluctuation-dissipation theorem relates the response functions to the structure factors as follows: $X^0_{AA}(Q) = \phi_A v_A S^0_{AA}(Q)/N_A$, etc. Recall that $S^0_{AA}(Q) = \langle \rho_A(-Q)\rho_A(Q) \rangle$. Moreover interacting response functions, $X_{AA}(Q) = \phi_A v_A S_{AA}(Q)/N_A = N_A \phi_A v_A P(\alpha_A N_A)$, etc., can be introduced through:

$$\langle \rho_A(Q) \rangle = - X_{AA}(Q)(U_A - U_B)/k_B T \qquad (6.2)$$

so that the externally applied potentials can be eliminated from these coupled equations. De Gennes' formula therefore relates the interacting response to the bare response functions:

$$1/X_{AA}(Q) = 1/X^0_{AA}(Q) + 1/X^0_{BB}(Q) - 2\chi_{AB}/v_0 \qquad (6.3)$$

where $v_0 = (v_A v_B)^{1/2}$ is the "lattice cell" volume and the Flory–Huggfins interaction "chi" parameter χ_{AB} has been defined as:

$$\chi_{AB} = W_{AB}/k_B T - (W_{AA} + W_{BB})/2k_B T . \qquad (6.4)$$

χ_{AB} appears here as a universal parameter. However, it was found experimentally to depend on a number of factors [30–33] such as temperature, molecular weight, composition, inter-monomer distance (and therefore on the scattering vector Q), isotopic constitution, tacticity, microstructure, etc. These dependencies are shortcomings of the crude RPA description. The scattered intensity (macroscopic cross section $d\Sigma(Q)/d\Omega$) is given by:

$$(a_A/v_A - a_B/v_B)^2/d\Sigma(Q)/d\Omega = 1/N_A \phi_A v_A P_G(\alpha_A N_A)$$
$$+ 1/N_B \phi_B v_B P_G(\alpha_B N_B) - 2\chi_{AB}/v_0 \qquad (6.5)$$

where we have used scattering length densities a_A/v_A, etc. This de Gennes formula has found wide use in polymer blends and has been instrumental in determining spinodal points (value of χ_{AB} for which the scattered intensity becomes infinite) and mapping out phase diagrams. Zimm's formula is recovered from the de Gennes formula by taking $N_A = N$, $N_B = 1$ and defining the excluded volume by: $v = v_p^2/v_s \phi_S - 2\chi_{PS} v_p^2/v_0$ (v is the excluded volume, v_s is the solvent molecule volume, v_0 is the cell volume, and V is the sample volume).

6.2 The Benoit–Akcasu Generalization

Benoit et al. [11–12] and Akcasu et al. [13–15] have extended de Gennes' formula to describe multicomponent polymer systems. Their results are reproduced in Appendix A in a matrix form (following Akcasu [13–15]). Consider a number of components (noted A, B, etc.) with degrees of polymerization N_A, etc., volume fractions ϕ_A, etc., monomer volumes v_A, etc. Some of these components could be block copolymers. Having one of the components (called "matrix" component) as a homopolymer simplifies the calculations. The main result is:

$$\mathbf{X}^{-1}(Q) = \mathbf{X}_0^{-1}(Q) + \mathbf{V}(Q) \tag{6.6a}$$

where $\mathbf{X}_0(Q)$ and $\mathbf{X}(Q)$ are the bare and interacting system response matrices respectively and $\mathbf{V}(Q)$ is an excluded volumes matrix (note that $v_0^2 \mathbf{V}(Q)$ and not $\mathbf{V}(Q)$ alone has the dimension of a volume). Equivalently:

$$\mathbf{X}(Q) = \mathbf{X}_0(Q) \cdot [\mathbf{I} + \mathbf{V}(Q) \cdot \mathbf{X}_0(Q)]^{-1} \tag{6.6b}$$

where I is the identity matrix. The macroscopic scattering cross section (scalar quantity) is obtained as:

$$d\Sigma(Q)/d\Omega = \mathbf{A}^T \cdot \mathbf{X}(Q) \cdot \mathbf{A} \tag{6.7}$$

where A is a column vector ($\mathbf{A}^T$ is the corresponding row vector) containing the scattering length densities of the various components (A, B, etc.) and that of the "matrix" (M) component; for example, $\mathbf{A}_A = (a_A/v_A - a_M/v_M)$. Appendix A describes multicomponent systems of homopolymers and copolymers. If only copolymers were present, then the "matrix" component would have to be one of the copolymer blocks; this situation involves more complicated expressions which are derived in Appendix B.

In the case of a compressible polymer mixture, the RPA yields the same result but with the potentials matrix W replacing the excluded volumes matrix V:

$$\mathbf{X} = \mathbf{X}_0 [\mathbf{I} + \mathbf{W} \cdot \mathbf{X}_0]^{-1}$$

Moreover the components of vector A change to become $\mathbf{A}_A = a_A/v_A$, etc. An Ornstein–Zernike (OZ) approach (referred to as the integral equation theory) describing multicomponent compressible polymer blend mixtures has been extensively investigated [35]. The multicomponent OZ equation relates the "direct" correlations matrix C and the "total" (i.e., direct and indirect) correlations matrix H as:

$$\mathbf{H} = \mathbf{X}_0 \mathbf{C} \mathbf{X}_0 + \mathbf{X}_0 \mathbf{C} \mathbf{H} . \tag{6.8a}$$

By definition, the structure factor is the sum of single-chain ("bare") $\mathbf{X_0}$ and interchain $\mathbf{H}$ structure factors: $\mathbf{X} = \mathbf{X_0} + \mathbf{H}$. The formal solution to the OZ equations:

$$\mathbf{H} = \mathbf{X_0} \mathbf{C} \mathbf{X_0} [\mathbf{I} - \mathbf{X_0} \mathbf{C}]^{-1} \tag{6.8b}$$

is identical to the RPA equation for compressible mixtures provided that a Mean Spherical Approximation (MSA) closure relation, $\mathbf{C} = -\mathbf{W}/k_B T$, is used for large intermonomer distances. The RPA approach considers the mean field potentials W's merely as parameters whereas realistic constraints (both for large and short distances) are used in order to solve the integral equations.

6.3 Three Component Flexible Homolymer Blend

The multicomponent RPA formalism [11–15] is applied, here, to a ternary incompressible mixture [11–16] of homopolymers (A, B, C). Assuming that component C is the "matrix" component, one is left with 2×2 matrices for components A and B:

$$X_{AA}^0(Q) = N_A \phi_A v_A P(\alpha_A N_A)$$

$$X_{BB}^0(Q) = N_B \phi_B v_B P(\alpha_B N_B)$$

$$X_{AB}^0(Q) = 0 \tag{6.9}$$

and the third component, C, enters only through:

$$V_{AA}(Q) = 1/X_{CC}^0(Q) - 2\chi_{AC}/v_0$$

$$V_{BB}(Q) = 1/X_{CC}^0(Q) - 2\chi_{BC}/v_0$$

$$V_{AB}(Q) = 1/X_{CC}^0(Q) + \chi_{AB}/v_0 - \chi_{AC}/v_0 - \chi_{BC}/v_0 \tag{6.10}$$

where the N's, ϕ's, and v's are the degrees of polymerization, volume fractions, and monomeric volumes respectively. As customary, v_0 is defined as the volume of the reference cell. The single-chain structure factors $P(\alpha N)$'s are taken to be Debye functions for flexible polymers:

$$P(\alpha_A N_A) = 2[\exp(-Q^2 R_{gA}^2) - 1 + Q^2 R_{gA}^2]/Q^4 R_{gA}^4, \tag{6.11}$$

and the radius of gyration is given in terms of the statistical length b_A as $R_{gA}^2 = N_A b_A^2/6$. Partial structure factors can be obtained as:

$$X_{AA}(Q) = X_{AA}^0(1 + V_{BB} X_{BB}^0)/[(1 + V_{AA} X_{AA}^0)(1 + V_{BB} X_{BB}^0)$$
$$- V_{AB}^2 X_{AA}^0 X_{BB}^0]$$

$$X_{BB}(Q) = X_{BB}^0(1 + V_{AA} X_{AA}^0)/[(1 + V_{AA} X_{AA}^0)(1 + V_{BB} X_{BB}^0)$$
$$- V_{AB}^2 X_{AA}^0 X_{BB}^0]$$

$$X_{AB}(Q) = -X_{AA}^0 V_{AB} X_{BB}^0/[(1 + V_{AA} X_{AA}^0)(1 + V_{BB} X_{BB}^0)$$
$$- V_{AB}^2 X_{AA}^0 X_{BB}^0] \tag{6.12}$$

where the Q-dependence has been dropped for convenience. Note that the spinodal point is reached when the denominator of the partial structure factors vanishes.

The neutron scattered intensity (macroscopic differential scattering cross section $d\Sigma(Q)/d\Omega$) is given by:

$$d\Sigma(Q)/d\Omega = (a_A/v_A - a_C/v_C)^2 X_{AA}(Q) + (a_B/v_B - a_C/v_C)^2 X_{BB}(Q)$$
$$+ 2(a_A/v_A - a_C/v_C)(a_B/v_B - a_C/v_C)X_{AB}(Q) \qquad (6.13)$$

where the a's are the monomeric scattering lengths for the different components. Note that, due to the incompressibility assumption, this result is independent of the contrast between components A and B. Note also that the three components (A, B, C) do not have to be linear chains; they could correspond to whatever chain architecture provided that the single-chain structure factors are known.

6.4 Blend Mixture of a Copolymer and a Homopolymer (Both Flexible)

We assume, now, that the three component blend considered in the previous section consists of a copolymer A/B (could be a diblock, triblock, etc, or an alternating copolymer) and a homopolymer C [11–15]. The notation and formalism of the previous section hold but now $X_{AB}^0(Q) \neq 0$ (note that $X_{AB}^0(Q)$ shows a peak in the scattering function). The partial structure factors become:

$$X_{AA}(Q) = \{X_{AA}^0(1 + V_{BA}X_{AB}^0 + V_{BB}X_{BB}^0)$$
$$- X_{AB}^0(V_{BA}X_{AA}^0 + V_{BB}X_{BA}^0)\}/\Delta$$
$$X_{BB}(Q) = \{X_{BB}^0(1 + V_{AB}X_{BA}^0 + V_{AA}X_{AA}^0)$$
$$- X_{BA}^0(V_{AB}X_{BB}^0 + V_{AA}X_{AB}^0)\}/\Delta$$
$$X_{AB}(Q) = \{ - X_{AA}^0(V_{AA}X_{AB}^0 + V_{AB}X_{BB}^0)$$
$$+ X_{AB}^0(1 + V_{AA}X_{AA}^0 + V_{AB}X_{BA}^0)\}/\Delta \qquad (6.14)$$

where:

$$\Delta = (1 + V_{AA}X_{AA}^0 + V_{AB}X_{BA}^0)(1 + V_{BA}X_{AB}^0 + V_{BB}X_{BB}^0)$$
$$- (V_{AA}X_{AB}^0 + V_{AB}X_{BB}^0)(V_{BA}X_{AA}^0 + V_{BB}X_{BA}^0). \qquad (6.15)$$

The V's and the bare structure factors $X_{AA}^0(Q)$ and $X_{BB}^0(Q)$ are still given by the same expressions as in the previous section, and:

$$X_{AB}^0(Q) = (\phi_A\phi_B v_A v_B)^{1/2} S_{AB}^0(Q)/(N_A N_B)^{1/2} \qquad (6.16)$$

and $S_{AB}^0(Q)$ are given in Sect. 3 for diblock, and alternating copolymers for example.

7 Multicomponent Blends of Stiff Homopolymers and Copolymers

7.1 General RPA Equations for Mixtures of Stiff Polymers

Consider a number n of stiff polymer components (here "stiff" is used to mean "semiflexible") and define orientation-dependent ideal and interacting response $(n \times n)$ matrices $X_0(Q, u, u')$ and $X(Q, u, u')$ respectively. In this case, orientational correlations have to be included in addition to the usual isotropic ones. Doi et al. [36–38] have developed the theory for solutions of stiff homopolymers. Their formalism is applied in Appendices C and D to multicomponent blend mixtures of stiff polymers without and with the incompressibility condition respectively. The interaction potentials comprise anisotropic (also called nematic) contributions as well as the usual isotropic ones:

$$W(u', u'') = W_0 - W_1(u'u' - I/3):(u''u'' - I/3) , \tag{7.1}$$

where u' and u'' represent the orientations of two test rods. The W_1 potential factors are the Maier–Saupe interaction parameters. The main result for the case of compressible stiff polymer mixtures (see Appendix C) is:

$$\begin{aligned} X = \{I + X_0 \cdot W_0 + (2/3)R_0^T \cdot W_1 \cdot M^{-1} \cdot R_0 \cdot W_0\}^{-1} \\ \times \{X_0 + (2/3)R_0^T \cdot W_1 \cdot M^{-1} \cdot R_0\} \end{aligned} \tag{7.2}$$

where the matrix $M = [I - (2/3)T_0 \cdot W_1]$ has been used, $k_B T$ dividing the potential parameters has been omitted for notational convenience, and the following orientational moments $(n \times n$ matrices) of the ideal structure factors have been defined:

$$X_0(Q) = \int du \int du' X_0(Q, u, u')$$

$$X(Q) = \int du \int du' X(Q, u, u')$$

$$R_0(Q) = (3/2)\int du \int du' X_0(Q, u, u')[(q \cdot u)^2 - 1/3] \tag{7.3}$$

$$R(Q) = (3/2)\int du \int du' X(Q, u, u')[(q \cdot u)^2 - 1/3]$$

$$T_0(Q) = (9/4)\int du \int du' X_0(Q, u, u')[(q \cdot u)^2 - 1/3][(q \cdot u')^2 - 1/3] .$$

where rq is the unit vector along Q. Although the incompressibility assumption is expected to be a reasonable one for flexible polymer mixtures, it is not known whether it could also be realistic for mixtures of flexible and rigid polymers. Making that assumption, the general result (for incompressible mixtures of stiff polymers) (see Appendix D) is:

$$\begin{aligned} X_{RR}^{-1} = P^T \cdot \{X_0 + (2/3)R_0^T \cdot W_1 \cdot M^{-1} \cdot R_0\}^{-1} \\ \times \{I + X_0 \cdot W_0 + (2/3)R_0^T \cdot W_1 \cdot M^{-1} \cdot R_0 \cdot W_0\} \cdot P \end{aligned} \tag{7.4}$$

where we have followed Akcasu [15] and introduced an $n \times (n - 1)$ matrix $\mathbf{P} = \mathrm{Col}[\mathbf{I}, -\mathbf{E}_R^T]$ ($\mathbf{I}$ is the $(n - 1) \times (n - 1)$ identity matrix and $\mathbf{E}_R^T$ is an $(n - 1)$ row vector comprising only ones) in order to apply the incompressibility constraint at the outset.

The isotropic-to-nematic transition is defined by the characteristic equation $\mathrm{Det}\{\mathbf{M}\} = 0$ (where Det represents the determinant of a matrix). If the Van der Waals interactions were "turned off" ($\mathbf{W}_0 = 0$) so that only nematic interactions are left, then $\mathbf{M}$ would be the denominator of $\mathbf{X}$ so that $\mathbf{X}$ would blow up for this condition ($\mathrm{Det}\{\mathbf{M}\} = 0$). Above certain critical values of $\mathbf{W}_1$'s the blend forms the nematic phase. As in the case of purely flexible mixtures, the spinodal condition is:

$$\mathrm{Det}\{\mathbf{I} + \mathbf{X}_0 \cdot \mathbf{W}_0 + (2/3)\mathbf{R}_0^T \cdot \mathbf{W}_1 \cdot \mathbf{M}^{-1} \cdot \mathbf{R}_0 \cdot \mathbf{W}_0\} = 0 . \tag{7.5}$$

Here also $k_B T = 1$ has been set. Since blend mixtures of completely rigid rods do not exist in the one-phase region, the approach described here, will be applied to mixtures of rigid and flexible polymers. The case of polymer solutions (results from Doi et al. [36–38]) can be recovered when one of the components is taken to be a solvent.

7.2 Binary Blend of a Flexible and a Rigid Rod Polymers

In the case of a binary incompressible mixture of stiff homopolymers (components are named A and B), the above equations simplify. Assuming that component A is flexible (freely-jointed chains) and B is rigid (rigid rod polymers) and imposing the incompressibility condition, the following result can be obtained:

$$X_{AA} = \{(2/3)R_{BB}^{02}W_{BB}^1 X_{AA}^0 + [1 - (2/3)T_{BB}^0 W_{BB}^1]X_{AA}^0 X_{BB}^0\}/$$
$$\{(2/3)R_{BB}^{02}W_{BB}^1[1 - 2\chi_{AB}X_{AA}^0]$$
$$+ (X_{AA}^0 + X_{BB}^0 - 2\chi_{AB}X_{AA}^0 X_{BB}^0)[1 - (2/3)T_{BB}^0 W_{BB}^1]\} \tag{7.6}$$

where $X_{AB}^0 = R_{AA}^0 = R_{AB}^0 = R_{BA}^0 = T_{AA}^0 = T_{AB}^0 = 0$ and the remaining ideal structure factors can readily be calculated.

$$X_{AA}^0(Q)/N_A \phi_A v_A = \{[1 + j_0(Qb_A)]N_A/[1 - j_0(Qb_A)]$$
$$- 2j_0(Qb_A)[1 - j_0^{N_A}(Qb_A)]/$$
$$[1 - j_0(Qb_A)]^2\}/N_A^2 \tag{7.7a}$$

$$X_{BB}^0(Q)/N_B \phi_B v_B = \int_0^1 dx\, j_0^2(Qb_B N_B x/2)$$
$$= 2[\cos(Qb_B N_B) - 1]/(Qb_B N_B)^2$$
$$+ 2Si(Qb_B N_B)/Qb_B N_B \tag{7.7b}$$

$$R_{BB}^0(Q)/N_B\phi_B v_B = (3/2)\int_0^1 dx\,(x^2 - 1/3)j_0^2(Qb_B N_B x/2)$$

$$= (3/2)\{2[5 - \cos(Qb_B N_B)]/3(Qb_B N_B)^2$$

$$- 2\sin(Qb_B N_B)/(Qb_B N_B)^3$$

$$- 2Si(Qb_B N_B)/3Qb_B N_B\} \qquad (7.7c)$$

$$T_{BB}^0(Q)/N_B\phi_B v_B = (9/4)\int_0^1 dx\,(x^2 - 1/3)^2 j_0^2(Qb_B N_B x/2)$$

$$= (9/4)\{ - 10/9(Qb_B N_B)^2 + \cos(Qb_B N_B)$$

$$\times [2/9(Qb_B N_B)^2 - 4/(Qb_B N_B)^4]$$

$$+ \sin(Qb_B N_B)[4/(Qb_B N_B)^5 - 2/3(Qb_B N_B)^3]$$

$$+ 2Si(Qb_B N_B)/9(Qb_B N_B)\} \qquad (7.7d)$$

where $j_0(X) = \sin(X)/X$ is the spherical Bessel function of order zero and $Si(X)$ is the sine integral function.

The isotropic-to-nematic transition is determined by the condition $[1 - (2/3)T_{BB}^0 W_{BB}^1/k_B T] = 0$ whereas the spinodal line is obtained when the denominator of X_{AA} is equal to zero. These conditions are evaluated in the thermodynamic limit $(Q = 0)$ in Fig. 7 for a Maier–Saupe interaction parameter $W_{BB}^1/k_B T = 0.4\chi_{AB}$ and for $N_A = 200$, $N_B = 800$, $v_A = v_B = 1$. When the volume fraction of component $A(\phi_A)$ is low, the isotropic-to-nematic phase transition is reached first whereas at high ϕ_A the spinodal line is reached first. In the second case, the macromolecules do not have a chance to orient themselves before the spinodal line is reached. This RPA approach is a generalization of the Doi et al. [36–38] results (that were developed for lyotropic polymer liquid crystals) to describe thermotropic polymer mixtures. Both approaches cannot, however,

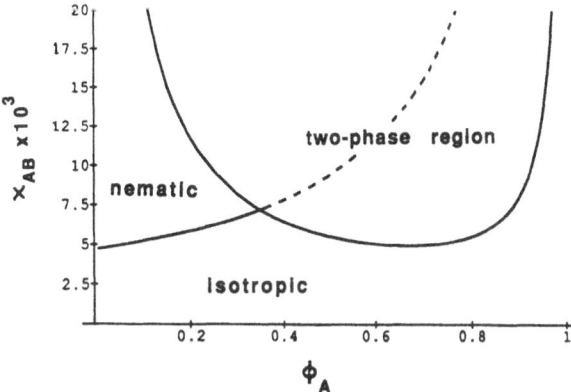

Fig. 7. Phase diagram for a binary blend mixture of a flexible (A component) and a rigid (B component) polymers with: $N_A = 200$, $N_B = 800$, $v_A = v_B = 1$, and $W_{BB}^1/k_B T\chi_{AB} = 0.4$ as predicted by the RPA

reproduce the narrow two-phase region between the isotropic and nematic phases as observed experimentally or as predicted by the lattice model approach [44]. Because the isotropic-to-nematic phase transition is a first order transition, the nematic phase can be obtained only after separation into two phases (one nematic and one isotropic) upon heating. The RPA formalism presented here is not valid beyond the spinodal line.

7.3 Binary Mixture of a Flexible and a Rigid Rod Diblock Copolymer

In the case of a diblock copolymer with flexible A blocks (freely jointed chains) and rigid B blocks (rigid rods), the intercomponent ideal structure factors are no longer equal to zero therefore leaving only: $R_{AA}^0 = R_{AB}^0 = T_{AA}^0 = T_{AB}^0 = 0$. The volume fractions become related to the molecular weights: $\phi_A = N_A v_A/(N_A v_A + N_B v_B)$, $\phi_B = 1 - \phi_A$. The general result in the matrix form reduces in this case to the following generalization of the Leibler formula [39] to include chain stiffness:

$$
\begin{aligned}
X_{AA} = & \{(2/3)W_{BB}^1[R_{BB}^{02}X_{AA}^0 - 2R_{BA}^0 R_{BB}^0 X_{AB}^0 + R_{BA}^{02}X_{BB}^0] \\
& + [1 - (2/3)T_{BB}^0 W_{BB}^1](-X_{AB}^{02} + X_{AA}^0 X_{BB}^0)\}/ \\
& \{(X_{AA}^0 + 2X_{AB}^0 + X_{BB}^0 + 2\chi_{AB}X_{AB}^{02} - 2\chi_{AB}X_{AA}^0 X_{BB}^0) \\
& \times [1 - (2/3)T_{BB}^0 W_{BB}^1] \\
& + (2/3)W_{BB}^1[R_{BA}^{02} + 2R_{BA}^0 R_{BB}^0 + R_{BB}^{02}] \\
& - (4/3)\chi_{AB}W_{BB}^1[R_{BB}^{02}X_{AA}^0 - 2R_{BA}^0 R_{BB}^0 X_{AB}^0 + R_{BA}^{02}X_{BB}^0]\} \quad (7.8)
\end{aligned}
$$

where the ideal structure factors X_{AA}^0, X_{BB}^0, R_{BB}^0, and T_{BB}^0 are given in the previous section and the remaining ones are given below:

$$
\begin{aligned}
& X_{AB}^0(Q)/(N_A\phi_A v_A N_B\phi_B v_B)^{1/2} \\
& = \{[1 - j_0(Qb_A)^{N_A}]/N_A[1 - j_0(Qb_A)]\} \\
& \quad \times \int_0^1 dx\, j_0(Qb_B N_B x/2)\cos(Qb_B N_B x/2) \\
& = \{[1 - j_0(Qb_A)^{N_A}]/N_A[1 - j_0(Qb_A)]\}\, Si(Qb_B N_B)/Qb_B N_B \quad (7.9a)
\end{aligned}
$$

$$
\begin{aligned}
& R_{BA}^0(Q)/(N_A\phi_A v_A N_B\phi_B v_B)^{1/2} \\
& = \{[1 - j_0(Qb_A)^{N_A}]/N_A[1 - j_0(Qb_A)]\} \\
& \quad \times \int_0^1 dx\, [3x^2 - 1/2]j_0(Qb_B N_B x/2)\cos(Qb_B N_B x/2) \\
& = \{[1 - j_0(Qb_A)^{N_A}]/N_A[1 - j_0(Qb_A)]\}(3/2)\{-\cos(Qb_B N_B)/ \\
& \quad (Qb_B N_B)^2 - Si(Qb_B N_B)/3(Qb_B N_B) \\
& \quad + \sin(Qb_B N_B)/(Qb_B N_B)^3\}\,. \quad (7.9b)
\end{aligned}
$$

These results agree with those reported by Holyst and Schick [40–41]. The structure factor $X_{AA}(Q)$ has been plotted in Fig. 8 using the following parameters: $N_A = 200$, $N_B = 800$, $v_A = v_B = 1$, $\chi_{AB} N = 19$ (where $N = N_A + N_B$), and for three values of the Maier–Saupe interaction parameter: $W_{BB}^1/k_B T \chi_{AB} = 0$, 0.4 and 0.6. Orientational ordering is seen to increase as the Maier–Saupe parameter increases. The location of the peaks in Figure 8 corresponds to $Q b_B N_B = 2\pi, 4\pi$ and depends only on the length of the rigid rods ($b_B N_B$). The peaks observed here are a characteristic of taking orientational moments of the structure factor for a rigid rod; these peaks appear even in the ideal (unperturbed) rigid rod case ($R_{BB^o}(Q)$ and $T_{BB^o}(Q)$, for example, show peaks at $Q b_B N_B = 2\pi, 4\pi$). Figure 9, on the other hand, represents the effect of varying the relative molecular weight of the flexible block with: $v_A = v_B = 1$, $\chi_{AB} N = 15$, $W_{BB}^1/k_B T \chi_{AB} = 0.4$ and N_A/N taking on three different values: $N_A/N = 0.2$, 0.7 and 0.8. The first curve corresponds to a point in the phase diagram which is closer to the isotropic-to-nematic phase transition line than to the isotropic-to-lamellar transition line. The other two curves correspond to points that are closer to the isotropic-to-lamellar spinodal line instead. The word "lamellar" is used to name the ordered phase even though this could have another morphology since the RPA cannot predict the symmetry of the ordered phase. Figure 9 shows that when the rigid rods get shorter, the number of peaks decreases (from two to one) because the first peak occurs at higher Q so that the higher order peaks are completely "damped" out. The sharpness of the peaks in Figure 8 points to the fact that the domain boundaries (in direct space) are

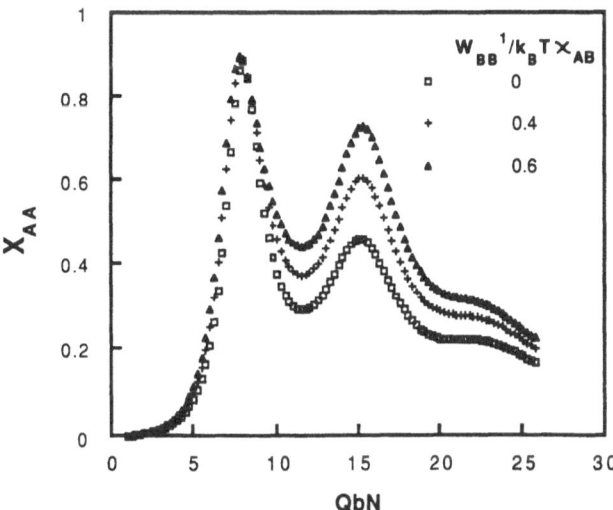

Fig. 8. Structure factor for a melt of diblock copolymers made of flexible freely-jointed (A component) and rigid (B component) blocks with: $N_A = 200$, $N_B = 800$, $v_A = v_B = 1$, $b_A = b_B = b$, and $\chi_{AB} N = 19$ (where $N = N_A + N_B$). The three curves correspond to $W_{BB}^1/k_B T \chi_{AB} = 0$, 0.4, and 0.6

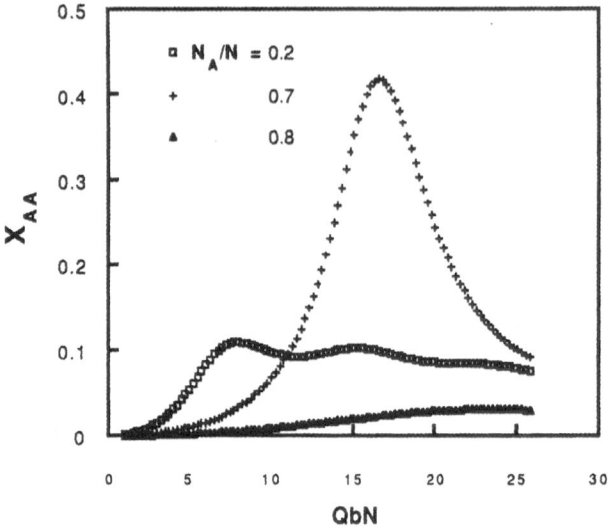

Fig. 9. Structure factor for a melt of diblock copolymers made of flexible freely-jointed (A component) and rigid (B component) blocks with: $\chi_{AB}N = 19$ (where $N = N_A + N_B$), $W^1_{BB}/k_B T\chi_{AB} = 0.4$, $v_A = v_B = 1$ and $b_A = b_B = b$. The three curves correspond to $N_A/N = 0.2$, 0.7, and 0.8 respectively

sharper in this system. When the rigid rods get shorter (Figure 9), this sharpness decreases leading to a regular sinusoidal profile for the density of rigid rods. Various phase diagrams for diblock copolymers have been investigated by Holyst and Schick [41]. Here also, it is emphasized that the RPA approach should be considered only for qualitative observations because it cannot predict the narrow two-phase channel between the nematic and isotropic regions which is due to non-mean field contributions.

Now that the RPA has been applied to various polymer mixtures, specific cases are considered in the following sections along with SANS data analyzed with arguments similar to the ones discussed in the previous sections.

8 Specific Examples

In the following sections, examples of SANS investigations from polymer systems are considered. Simple cases involving polymer solutions and blends are described in order to demonstrate the modeling approaches discussed here. These examples have been borrowed from my recent work in collaboration with other scientists at the National Institute of Standards and Technology.

8.1 Semidilute Solution of Deuterated Polystyrene in Dioctyl Phthalate

A semidilute solution [42] of high molecular weight deuterated polystyrene ($M_w = 1.95 \times 10^6$ g/mole, $M_w/M_n = 1.64$) in dioctyl phthalate (DOP) at a volume fraction of 2.83% of polystyrene was measured by SANS at room temperature. A characteristic intensity behavior $I(Q)$ was obtained after data correction (solvent incoherent scattering, empty cell scattering and usual background corrections, etc.) and was circularly averaged. The reduced $I(Q)$ data was then fitted to the following form:

$$(a_p/v_p - a_s/v_s)^2/I(Q) = 1/N\phi_p v_p P_G(\alpha N) + 1/\phi_s v_s - 2\chi_{ps}/v_0 \qquad (8.1)$$

where the same notation is being used: degree of polymerization: $N = 18750$, polymer volume fraction: $\phi_p = 2.83\%$, molar volumes: $v_p = 100$ cm^3, $v_s = 398$ cm^3, contrast factor: $(a_p/v_p - a_s/v_s)^2 N_{av} = 5.57 \times 10^{-3}$ mole/cm^4, Debye function:

$$P_G(\alpha N) = 2[\exp(-\alpha N) - 1 + \alpha N]/(\alpha N)^2 \qquad (8.2)$$

with fully swollen radius of gyration:

$$\alpha N = Q^2 R_g^2 = Q^2 b^2/(2v + 1)(2v + 2) \qquad (8.3)$$

and $v = 0.6$ (good solvent conditions). Results of the fit were: segment length: $b = 0.725$ nm and interaction parameter: $\chi_{ps}/v_0 = 9.68 \times 10^{-4}$ mole/cm^3. The spinodal value is $\chi_{ps}/v_0 = 1/2N\phi_p v_p + 1/2\phi_s v_s = 1.297 \times 10^{-3}$ mole/cm^3. The experimentally obtained and the calculated intensities are shown in Fig. 10 for comparison.

8.2 Binary Blend of Deuterated Polystyrene and Poly(vinyl methyl ether)

Consider a binary polymer blend [43] of deuterated polystyrene, PSD, ($M_w = 1.95 \times 10^5$ g/mole, $M_w/M_n = 1.02$) and poly(vinyl methyl ether), PVME, ($M_w = 1.59 \times 10^5$ g/mole, $M_w/M_n = 1.3$) with a composition of 48.4% PSD (volume fraction). SANS data were taken at various temperatures ranging from ambient to 160 °C. De Gennes's RPA formula:

$$(a_A/v_A - a_B/v_B)^2/d\Sigma(Q)/d\Omega = 1/N_A\phi_A v_A P_G(\alpha_A N)$$
$$+ 1/N_B\phi_B v_B P_G(\alpha_B N) - 2\chi_{AB}/v_0 \qquad (8.4)$$

was used to fit the reduced data (Fig. 11) with A as the PSD component and B as the PVME component and with: $N_A = 1741$, $N_B = 2741$, $v_A = 100$ cm^3/mole, $v_B = 55.4$ cm^3/mole, $(a_A/v_A - a_B/v_B)^2 N_{av} = 6.07 \times 10^{-3}$ mole/cm^4, $\phi_A = 48.4\%$, etc. Results of the fits were: $b_{PSD} = 0.8$ nm, $b_{PVME} = 0.6$ nm and:

$$\chi_{PSD/PVME}/v_0 = 9.73 \times 10^{-4} - 0.416/T, \quad \text{(in mole/cm}^3\text{)} \qquad (8.5)$$

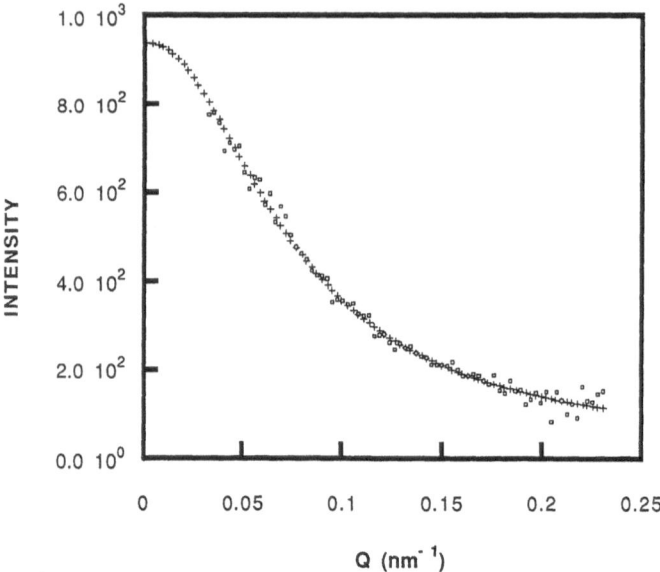

Fig. 10. SANS from deuterated polystyrene ($M_w = 1.95 \times 10^6$ g/mol) in dioctyl phthalate solution (3% polymer weight fraction). Experimental data (arbitrary units) and results of the fit to the inverse Zimm formula are plotted

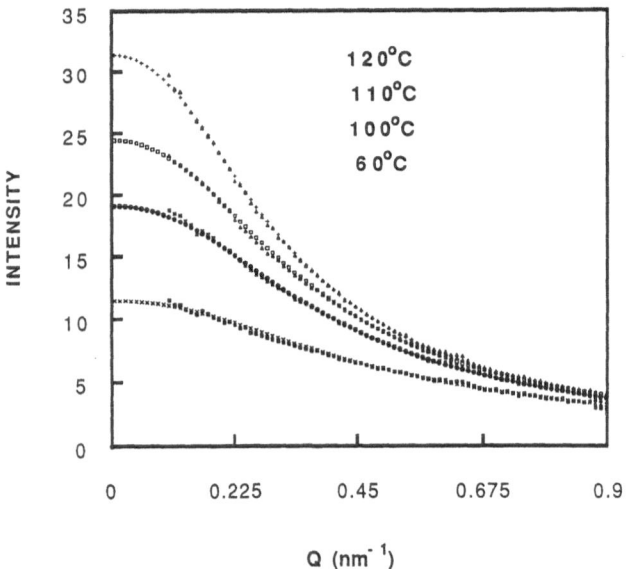

Fig. 11. SANS from deuterated polystyrene/poly(vinyl methyl ether) at equal compositions (50%/50% weight fractions). Experimental data (macroscopic cross section) and fits to the de Gennes formula are plotted for four temperatures: 60 °C, 100 °C, 110 °C and 120 °C (from bottom to top)

where T is the absolute temperature (Fig. 12). The PSD/PVME blend is a lower critical solution temperature (LCST) system and shows spinodal decomposition for $\chi_{PSD/PVME} = 1.23 \times 10^{-5}$ mole/cm^3 (value which makes the right-hand side of de Gennes's equation equal to zero); i.e., at 160 °C. These results agree with previously reported values [31] for this system.

8.3 Ternary Blend of Deuterated Polystyrene/ Poly(vinyl methyl ether)/Protonated Polystyrene; The High Concentration Method

Consider a ternary homopolymer blend mixture of PSD, PVME and protonated polystyrene (PSH). PSD and PVME have the same molecular weights as in the previous section and for PSH: $M_w = 1.90 \times 10^5$ g/mole, $M_w/M_n = 1.04$. The extra parameters needed to describe the blend are: $N_{PSH} = 1827$, $v_{PSH} = 100$ cm^3/mole, $(a_{PSH}/v_{PSH} - a_{PVME}/v_{PVME})^2 N_{av} = 1.79 \times 10^{-4}$ mole/cm^4. Three compositions corresponding to the same PVME volume fraction were measured by SANS:

> sample 1: 48.4%/51.6%/0%   PSD/PVME/PSH
>
> sample 2: 36%/51.1%/12.9%   PSD/PVME/PSH
>
> sample 3: 23.8%/50.6%/25.6%   PSD/PVME/PSH .

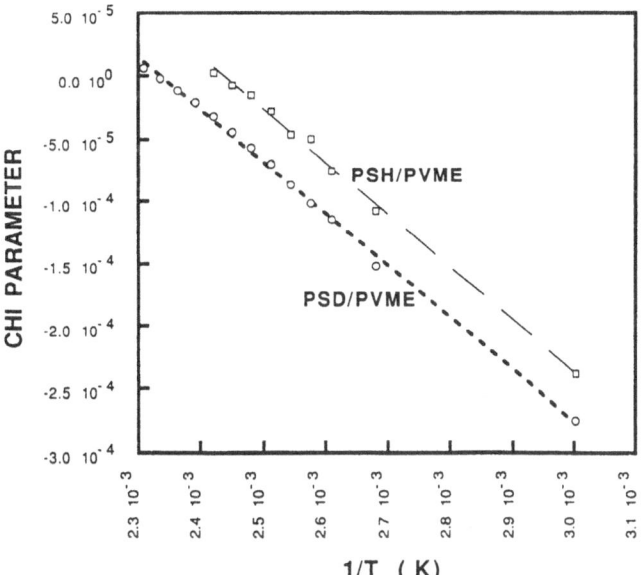

Fig. 12. Flory-Huggins χ/v_0 parameters for deuterated polystyrene/poly(vinyl methyl ether) and protonated polystyrene/poly(vinyl methyl ether) interactions. The first one was obtained from binary (PSD/PVME) mixtures (50%/50% weight fraction) and the second one from ternary (PSD/PVME/PSH) blend mixtures (23.8%/25.6%/50.6%)

The high concentration method is used, here, to extract single-chain, interchain and total scattering structure facrtors. The scattering cross section for each sample is given by:

$$d\Sigma(Q)/d\Omega = (a_C/v_P - a_A/v_P)^2 [\phi_A \phi_C / \phi_P^2] N \phi_P v_P P_S(Q)$$
$$+ [(a_A/v_P - a_B/v_B)\phi_A/\phi_P$$
$$+ (a_C/v_P - a_B/v_B)\phi_C/\phi_P]^2 N \phi_P v_P P_T(Q), \qquad (8.6)$$

where $N_A = N_C = N$, $v_A = v_C = v_P$, $\phi_A + \phi_C = \phi_P$. $P_S(Q)$, $P_T(Q)$ and the inter-chain structure factor $(P_I(Q) = [P_T(Q) - P_S(Q)]/\phi_P)$ can be extracted by combining data taken for pairs of samples (sample pairs 1–2, 2–3 and 1–3) with the proper weighing factors. The results are shown in Fig. 13. A number of observations can be made. For instance, the various structure factors extracted from different pairs of samples are slightly different due to the fact that the high concentration method assumptions hold only approximately (conformations may change from sample to sample, $P_{AA}^I(Q)$, $P_{AC}^I(Q)$ and $P_{CC}^I(Q)$ may not be identical, the system may not be completely incompressible, etc.). Moreover, the interchain structure factors $P_I(Q)$ are negative as they should. Recall that, for example,

$$P_{AC}^I(Q) = -[V/Nv_P] \int d^3R [1 - g_{AC}(R)] \exp[-iQ \cdot R] \qquad (8.7)$$

where $g_{AC}(R)$ is the pair distribution function representing monomer packing and R is the inter-monomers distance. $P_{AC}^I(Q)$ is obviously a very complicated

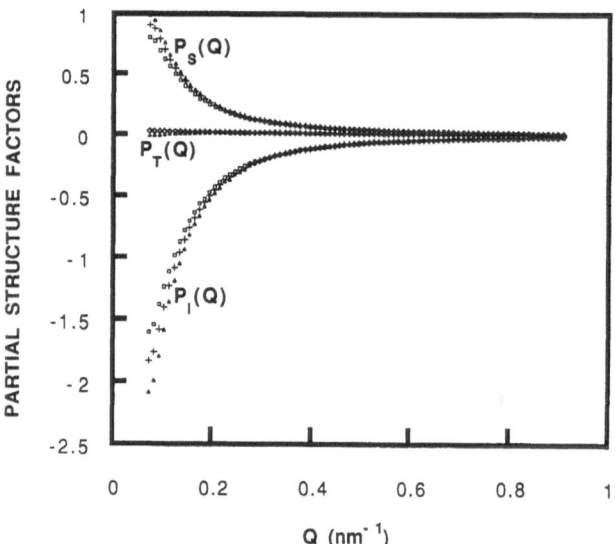

Fig. 13. Single-chain $P_S(Q)$, interchain $P_I(Q)$ and total $P_T(Q)$ structure factors for a blend mixture of deuterated and protonated polystyrene (PSD, PSH) in poly(vinyl methy ether) (PVME). The total PVME fraction was 51% and the PSD fractions were varied from 49% to 24%. The three curves correspond to sample pairs 1–2, 2–3 and 1–3 in each case

quantity to model directly in blend systems. It contains information about the various monomer–monomer interaction potentials, about monomer packing, and about overall chain conformations. For instance, $P_{AC}^l(Q)$ depends not only on AC interactions (represented by the χ_{AC} "chi" parameter) but also on the other ones (χ_{AB}, χ_{BC}). In order to investigate slight changes in the "chi" parameter due to deuteration, the RPA is used, instead, to analyze the SANS data from one of the ternary blends in the next section.

8.4 Ternary Blend of Deuterated Polystyrene/ Poly(vinyl methyl ether)/Protonated Polystyrene; The RPA Method

Consider one of the ternary blend mixtures described in the previous section. Data from sample 3 were taken from room temperature to 160 °C and are analyzed [43], using the RPA formalism for a ternary blend. The three components are called: A: PSD, B: PVME, C: PSH. Also, temperature dependencies for the two known chi parameters ($\chi_{PSD/PVME}/v_0$ and $\chi_{PSD/PSH}/v_0$) were assumed [31, 32]:

$$\chi_{PSD/PVME}/v_0 = 9.73 \times 10^{-4} - 0.416/T$$

$$\chi_{PSD/PSH}/v_0 = -2.9 \times 10^{-6} + 0.0020/T .$$

(8.8)

Results of the fits showed that $\chi_{PSH/PVME}/v_0$ has the following temperature dependence:

$$\chi_{PSH/PVME}/v_0 = 10.6 \times 10^{-4} - 0.436/T .$$

(8.9)

This dependence suggests a spinodal temperature for the PSH/PVME system of 140 °C which is in agreement with cloud point measurements. Figure 12 shows that the temperature dependence of $\chi_{PSD/PVME}$ and $\chi_{PSH/PVME}$ are parallel indicating that deuteration brings about a uniform shift in the spinodal temperature. This result, however, may not hold for other compositions.

9 Discussion

Monodisperse polymer blocks have been assumed all along. Polydispersity effects could be introduced in an ad-hoc fashion by assuming a molecular weight distribution (such as the Zimm–Schultz):

$$W(n) = [a/\langle n \rangle \Gamma(a + 1)](a/\langle n \rangle)^a \exp(-an/\langle n \rangle)$$

(9.1)

where $\langle n \rangle$ is the number average degree of polymerization, a is a measure of the polydispersity, and $\Gamma(a + 1)$ is the gamma function. The weight average degree of polymerization is

$$n_w = \langle n^2 \rangle / \langle n \rangle = \langle n \rangle (a + 1)/a \qquad (9.2)$$

so that the degree of polydispersity is $\varepsilon = (n_w - \langle n \rangle)/\langle n \rangle = 1/a$. Averaging over this distribution has the following effect:

$$\langle \exp(-\alpha n) \rangle = (1 + \alpha \langle n \rangle \varepsilon)^{-1/\varepsilon} . \qquad (9.3)$$

and the Debye function becomes:

$$\langle P(\alpha n) \rangle = 2[(1 + \alpha \langle n \rangle \varepsilon)^{-1/\varepsilon} - 1 + \alpha \langle n \rangle \varepsilon]/(1 + \varepsilon)\alpha^2 \langle n \rangle^2 . \qquad (9.4)$$

Note that the same symbol $\langle \cdots \rangle$ was used before to denote a conformational average while here it denotes a molecular weight average. Averaging over molecular weight distributions cannot always be done exactly. In cases where the degree of polymerization (n) appears in complicated expressions, "preaveraging" approximations ($\langle f(n)g(n) \rangle \sim \langle f(n) \rangle \langle g(n) \rangle$) have to be resorted to. Random copolymers, for instance, could be modeled as regularly alternating copolymers with polydisperse blocks [22].

With the advent of "formula manipulation" computer programs that perform analytical manipulations (Mathematica [45] was extensively used by this author), lengthy results for the various structure factors for multicomponent polymer mixtures are readily obtained and can form the basis of FORTRAN codes that can be used to fit neutron scattering data. This approach is preferable to performing the matrix inversions numerically because it involves an initial careful setting up of the general RPA formulas only. Also analytical functional forms are preferable to direct numerical calculations (matrix inversions, etc.) in least-squares fitting where these functions are evaluated thousands of times. This approach which consists in handling very complicated analytical forms to fit the data could be referred to as "semi-analytical".

The RPA theory works surprisingly well for homogeneous flexible polymer systems considering the crude approximations involved (mean field and linear response). It works better for concentrated solutions, melts and blends. It is also often used for dilute and semidilute solutions. It, however, breaks down in non-homogeneous (such as phase decomposing) systems and close to the critical point where equivalent tools are not yet available. Renormalization group theory, on the other hand, works well in dilute and semidilute polymer solutions because it can account for loop interactions within one chain and for higher level interchain (multiple) contacts; but breaks down in melts and blends. Moreover, scaling arguments are available for the estimation of exponents in polymer solutions. Scaling theory can be tested using neutron scattering (log-log plots); however, it cannot predict prefactors and cannot, therefore, be used to directly fit scattering data. The RPA itself has its own drawbacks; it cannot, for example, describe compact polymer systems (such as stars with a large number of arms or starburst dendrimers) in the semidilute region where these show local "liquid-like" ordering (appearance of an interparticle peak in the scattering function). It can however, describe these systems in the concentrated region (where chains interpenetrate). The effect of Coulomb interactions (in polyelectrolytes for example) or dipolar interactions (in ionomers for example) have not been discussed here. These interactions (as well as hydrogen bonding) increase the long range order therefore enhancing density fluctuations and making the mean field assumption (inherent in the RPA) a weaker approximation.

The Flory–Huggins "chi" parameter was introduced as a universal parameter. It has been found to depend on a number of experimental variables (molecular weight, composition, microstructure, tacticity, intermonomer distance, etc.) so that it has become an empirical parameter used to disguise our inadequacies in the knowledge of polymer systems. Modeling efforts leading to a better microscopic understanding of the origin of monomer–monomer and/or monomer/solvent interactions are needed. Van der Waals interactions between atoms (represented by electronic polarizabilities) could form the basis of such efforts. Other outstanding topics that require modeling efforts include: structure factor for polymer networks, effects of shear on the scattering function (the RPA may not be appropriate to use in the presence of shear), simple non-mean field treatments of polymer solutions that could be used to fit scattering data, prediction of phase diagrams for stiff copolymer mixtures, etc.

When the SANS technique was introduced, twenty years ago, data analysis consisted mainly of observations of relative trends of parameters (such as the radius of gyration) obtained by performing standard plots (such as the Guinier plot) and the SANS technique was not "exciting". However, modeling tools such as the ones described here have made SANS a very precise technique that is rich in information making it an increasingly used method (despite the fact that SANS spectra look "dull" compared to other spectra) with applications ranging from cutting edge science all the way to applied routine characterization. Newly introduced methods (such as the High Concentration Method or the Random Phase Approximation) have brought renewed interest in this technique. A recent literature search by this author of the Chemical Abstracts database based on the two keywords "neutron" and "polymer" came up with 480 articles that were published between 1980 and 1990, 424 of which used the SANS technique. New experimental methods both in the synthesis of ingenius polymer structures as well as in making judicious sample environments (such as the shear cell) are opening up new horizons to the SANS technique. Shearing, for instance, is making this "equilibrium" tool valuable for the investigation of rheology problems as well. Kinetic measurements (time slicing of the data) is allowing dynamic studies of chain conformations and concentration fluctuations to be made. Phase separation of blends occurs so rapidly that the peak in the scattering function (characterizing the later stages of spinodal decomposition) forms and disappears into the very low-Q region rapidly. A few systems (such as polycarbonate/polymethyl metacrylate) are characterized by slow phase separation kinetics which coupled with the availability of low-Q instruments and of time slicing will permit the investigation of the intermediate stages of spinodal decomposition. Theoretical back up is available both in early and late stages and predictions are being tested using light scattering.

It is unfortunate that research in the area of polymer solutions has been deserted during the past ten years in favor of blend work. With the advent of shear cells, it is expected that research in polymer solutions will become "fashionable" again. For instance, the phenomenon of shear-induced apparent demixing of high molecular weight polystyrene in semidilute solutions (in DOP for example) is not understood. Kinetics measurements will hopefully permit a close monitoring of the remixing effect after shear cessation as well.

The RPA formalism has been generalized to describe mixtures of stiff and flexible polymers. Eventhough this generalization reproduces the main expected features (isotropic-to-nematic transition for example), it cannot reproduce experimental phase diagrams observed in lyotropic liquid crystals for instance, whereby a narrow two-phase region (channel) is observed between the isotropic and nematic phases. The approach described here is a mere generalization of the Doi et al. approach [36–38] to describe thermotropic polymer liquid crystal mixtures. Most SANS measurements on liquid crystalline polymers have focussed on the investigation of chain conformations in oriented systems. There is a need for data in the isotropic region which cannot be reached in most systems. However, with slightly hydrogen bonded blends of flexible and rigid polymers, there is hope for extracting both the Flory-Huggins and the Maier–Saupe interaction parameters and investigating their dependences on temperature, molecular weight, and composition.

10 Appendices

Appendix A: Multicomponent Random Phase Approximation for Homopolymer and Copolymer Mixtures

Consider a polymer system consisting of n components. These could be homopolymer mixtures or homopolymer and copolymer mixtures. In order to simplify the calculations, we consider that at least one of the components (that we call "matrix" component) is a homopolymer. The formalism presented here is a straightforward extension of the two-component case in an n-vector and $n \times n$ matrix notation [13]:

$$\langle \rho(Q) \rangle = - X_0(Q)[U/k_B T + (W/k_B T) \cdot \langle \rho(Q) \rangle + \lambda E] \qquad (A.1a)$$

$$\langle \rho(Q) \rangle = - X(Q) U/k_B T \qquad (A.1b)$$

$$E^T \cdot \langle \rho(Q) \rangle = 0 \qquad (A.1c)$$

where $\langle \rho(Q) \rangle$ and U are the vector density and externally applied potentials, $X_0(Q)$ and $X(Q)$ are the bare and interacting system structure factor matrices, W is the monomer-monomer interaction potentials matrix, E is a vector with all components equal to unity and λ is a Lagrange multiplier introduced to help impose the incompressibility constraint (incompressibility equation: $E^T \cdot \langle \rho(Q) \rangle$

= 0). The idea is to isolate the "matrix" component (denoted component M) from the "rest" of the blend (denoted R). Matrix $X_0(Q)$ is formed of a scalar part $X_{MM}^0(Q)$, a vector part $X_{MR}^0(Q)$ and a matrix part $X_{RR}^0(Q)$ and similarly for W. Assuming that no copolymer is shared between the "matrix" component and the remaining $(n-1)$ components imposes $X_{MR}^0(Q) = 0$ therefore simplifying the calculations. With the M–R separation, the RPA equations become:

$$\langle \rho_M \rangle = - X_{MM}^0 U_M/k_B T - X_{MR}^0 U_R/k_B T - (X_0 \cdot W)_{MM} \langle \rho_M \rangle/k_B T$$

$$- (X_0 \cdot W)_{MR} \langle \rho_R \rangle/k_B T - \lambda(X_{MM}^0 + X_{MR}^0 \cdot E_R) \,, \qquad (A.2a)$$

$$\langle \rho_R \rangle = - X_{RM}^0 U_M/k_B T - X_{RR}^0 U_R/k_B T - (X_0 \cdot W)_{RM} \langle \rho_M \rangle/k_B T$$

$$- (X_0 \cdot W)_{RR} \langle \rho_R \rangle/k_B T - \lambda(X_{RM}^0 + X_{RR}^0 \cdot E_R) \,, \qquad (A.2b)$$

$$\langle \rho_M \rangle = - X_{MM} U_M/k_B T - X_{MR} U_R/k_B T \qquad (A.2c)$$

$$\langle \rho_R \rangle = - X_{RM} U_M/k_B T - X_{RR} U_R/k_B T \qquad (A.2d)$$

$$\langle \rho_M \rangle + E_R^T \cdot \langle \rho_R \rangle = 0 \qquad (A.2e)$$

where E_R is an $(n-1)$-vector with all terms equal to unity and E_R^T denotes its transpose. Extracting λ from Eq. (A.2a) and replacing it in Eq. (A.2b) yields an equation for $\langle \rho_R \rangle$, which with the help of Eq. (A.2e) becomes:

$$\langle \rho_R \rangle = - X_{RR}^0 \{ U_R/k_B T + (- W_{RM} \cdot E_R^T + W_{RR}) \cdot \langle \rho_R \rangle/k_B T$$

$$- E_R(X_{MM}^0)^{-1} [- E_R^T + X_{MM}^0(- W_{MM} E_R^T + W_{MR})/k_B T] \langle \rho_R \rangle \}$$

$$+ (X_{RR}^0 \cdot E_R) U_M/k_B T \qquad (A.3a)$$

$$\langle \rho_R \rangle = - X_{RM} U_M/k_B T - X_{RR} U_R/k_B T \,. \qquad (A.3b)$$

Defining an excluded volumes matrix:

$$V = - W_{RM} E_R^T/k_B T + W_{RR}/k_B T + E_R E_R^T/X_{MM}^0$$

$$+ E_R E_R^T W_{MM}/k_B T - E_R W_{RM}^T/k_B T \qquad (A.4)$$

the RPA equations become (I is the identity matrix):

$$X_{RR}(Q) = X_{RR}^0(Q) \cdot [I + V(Q).X_{RR}^0(Q)]^{-1} \qquad (A.5)$$

or if one drops the RR indices (in another form):

$$X^{-1}(Q) = X_0^{-1}(Q) + V(Q) \,. \qquad (A.6)$$

Note that without the incompressibility assumption (and without using the Lagrange multiplier λ), one would arrive to an identical equation for $n \times n$ matrices but with W replacing $V(Q)$. Note also that $V(Q)$ has the dimension of inverse volume.

This generalization of de Gennes' formula to multicomponent homopolymer and copolymer blends can describe a wide variety of situations. After a slight generalization (described in Appendix B) it can also, describe the case of pure copolymer mixtures whereby a copolymer has to be shared between M and R. Using Appendix A, one could obtain this limit (pure copolymer mixtures) by assuming a fictitious homopolymer (say component Z) in the mixture to be the matrix component, then taking the limit $\phi_Z \to 0$. This method is cumbersome because of the fact that the Z component appears in the denominator of the various terms of $V(Q)$ which diverge if proper care is not taken. Moreover, this procedure involves inverting $(n + 1) \times (n + 1)$ matrices for an n-component problem. The general situation [15] where a copolymer is shared between R and M is discussed in Appendix B.

Appendix B: Multicomponent Random Phase Approximation for Pure Copolymer Mixtures

In the case where $X_{MR}^0(Q) \neq 0$; i.e., if a copolymer is shared between R and M, we follow Akcasu [15] and introduce an $n \times (n - 1)$ matrix $P = Col[I, -E_R^T]$ where I is the $(n - 1) \times (n - 1)$ identity matrix and E_R^T is an $(n - 1)$ row vector comprising only ones. Using the P matrix, the incompressibility statement becomes: $\langle \rho \rangle = P \langle \rho_R \rangle$ and Eq. (A.1b) becomes: $X_{RR}^{-1} \langle \rho_R \rangle = -P^T U/k_B T$. Multiplying:

$$X_0^{-1}\{I + X_0 \cdot W_0/k_B T\} \cdot \langle \rho \rangle = -[U/k_B T + \lambda E] \tag{B.1}$$

on the left by P^T and using the fact that $P^T \cdot E = 0$, one can eliminate the potentials U and obtain:

$$X_{RR}^{-1} = P^T \cdot X_0^{-1}\{I + X_0 \cdot W_0/k_B T\} \cdot P . \tag{B.2}$$

This is a general result that states that the incompressibility constraint can be applied by sandwiching the $n \times n$ matrix $X^{-1}(Q)$ (obtained for compressible mixtures) between P^T and P in order to obtain the $(n - 1) \times (n - 1)$ matrix of structure factors $X_{RR}^{-1}(Q)$ for the remaining components. For example in a binary blend mixture of flexible polymers, the sandwiching procedure has the effect of adding the diagonal elements and subtracting the off-diagonal elements; i.e., $P^T W_0 P = (W_{AA}^0 + W_{BB}^0 - W_{AB}^0 - W_{BA}^0)/k_B T = -2\chi_{AB}$ which defines the Flory–Huggins interaction parameter χ_{AB} in terms of the interaction potentials W_0's. Of course, the approaches of Appendices A and B agree in common cases (for example in the case of a copolymer A–B and homopolymer C mixture).

Appendix C: Compressible Binary Blend Mixture of Stiff Polymers

We consider a polymer system consisting of n kinds of stiff polymers, and use the matrix notation approach introduced by Akcasu [13–15]. Some of these components could be copolymers. Component I has a degree of polymerizations N_I, volume fraction ϕ_I, monomer volume v_I, and segment size b_I. For stiff polymers, the averaged fluctuating density is defined as:

$$\langle \rho(Q, u) \rangle = \sum_{\alpha i} \langle \exp(-iQ \cdot r_{\alpha i})\delta(u - u_{\alpha i}) \rangle \tag{C.1}$$

where monomer i in polymer α is located at position $r_{\alpha i}$ and is oriented along direction $u_{\alpha i}$ and where $\langle \cdots \rangle$ represents an average over conformations (i.e., over distributions of $r_{\alpha i}$ and $u_{\alpha i}$).

Following the standard RPA formalism, we define externally applied (weakly perturbing) potentials U (U is an n-component vector that can depend on Q but not on monomer orientations) and inter-segment potentials $W(u, u')$ (n × n matrix) where u and u' represent the directions of two test segments. Within the mean field approach, the RPA equations give the mean response of the averaged densities $\langle \rho(Q, u) \rangle$ ($\langle \rho \rangle$ is an n-vector) in terms of the response functions for the bare system $X_0(Q, u, u')$ (n × n matrix) and for the interacting system $X(Q, u, u')$. In this matrix notation approach, bold face characters are used to represent n-vectors, n × n matrices as well as three-dimensional cartesian vectors such as direction u. The RPA equations in the matrix form are:

$$\langle \rho(Q, u) \rangle = - \int du' X_0(Q, u, u')$$
$$[U/k_B T + \int du'' W(u', u'') \langle \rho(Q, u'') \rangle / k_B T] \qquad (C.2a)$$

along with:

$$\langle \rho(Q, u) \rangle = - \int du' X(Q, u, u') U / k_B T . \qquad (C.2b)$$

Note that these two sets of equations can be combined to give:

$$X(Q, u, u') = X_0(Q, u, u') - \int du_1 du_2 X_0(Q, u, u_1)$$
$$\times W(u_1, u_2) X(Q, u_2, u') / k_B T \qquad (C.3)$$

which are the general RPA integral equations for compressible blend mixtures. Note that these equations are similar to the Ornstein–Zernicke relations [35]. Since the inter-segment interactions become weaker when two test segments are parallel to each other, the interaction potentials are taken to be proportional to $\sin(\alpha)$ where α is the angle between the two test segments ($\sin(\alpha) = |u' \times u''|$ where x is the vectorial product). In order to proceed further, Doi et. al. [36–38] assumed the following expansion:

$$\sin(\alpha) = (\pi/4)\{1 - (15/16)(u'u' - I/3):(u''u'' - I/3) + \cdots \} \qquad (C.4)$$

where $u'u'$ represents a second rank tensor, I is the second rank unity tensor and the column (:) represents the scalar product of two second rank tensors. Neglecting higher order terms effectively decouples the u' and u'' integrations in the RPA equations therefore making calculations tractable analytically. The interaction potentials can, therefore, be assumed to be:

$$W(u_1, u_2) = W_0 - W_1 (u_1 u_1 - I/3):(u_2 u_2 - I/3), \qquad (C.5)$$

W_0 and W_1 (n × n matrices) are assumed, here, to contain unknown potential parameters (Note that Doi et al. [36–38] relate their scalar counterparts through: $W_1/W_0 = 15/16$ for rigid rods). The W_1 potential factors are the

Maier–Saupe interaction parameters. The $\mathbf{u}_1$ and $\mathbf{u}_2$ integrations become:

$$\int d\mathbf{u}_1 \int d\mathbf{u}_2 \, X_0(Q, \mathbf{u}, \mathbf{u}_1) W(\mathbf{u}_1, \mathbf{u}_2) X(Q, \mathbf{u}_2, \mathbf{u}'')$$
$$= \{\textstyle\int d\mathbf{u}_1 X_0(Q, \mathbf{u}, \mathbf{u}_1)\} W_0 \{\textstyle\int d\mathbf{u}_2 X(Q, \mathbf{u}_2, \mathbf{u}'')\}$$
$$- \{\textstyle\int d\mathbf{u}_1 X_0(Q, \mathbf{u}, \mathbf{u}_1)(\mathbf{u}_1 \mathbf{u}_1 - \mathbf{I}/3)\}:$$
$$W_1 : (\textstyle\int d\mathbf{u}_2 X(Q, \mathbf{u}_2, \mathbf{u}'')(\mathbf{u}_2 \mathbf{u}_2 - \mathbf{I}/3)\} . \tag{C.6}$$

We use the following identity:

$$\int d\mathbf{u}' X_0(Q, \mathbf{u}, \mathbf{u}')[\mathbf{u}'\mathbf{u}' - \mathbf{I}/3] = (3/2)[\mathbf{qq} - \mathbf{I}/3]$$
$$\times \int d\mathbf{u}' X_0(Q, \mathbf{u}, \mathbf{u}')[(\mathbf{q} \cdot \mathbf{u}')^2 - 1/3] \tag{C.7}$$

(where the unit vector $\mathbf{q} = \mathbf{Q}/|\mathbf{Q}|$ has been used to represent the longitudinal direction) and define the following orientational moments ($n \times n$ matrices):

$$X_0(Q) = \int d\mathbf{u} \int d\mathbf{u}' X_0(Q, \mathbf{u}, \mathbf{u}')$$
$$X(Q) = \int d\mathbf{u} \int d\mathbf{u}' X(Q, \mathbf{u}, \mathbf{u}')$$
$$R_0(Q) = (3/2) \int d\mathbf{u} \int d\mathbf{u}' X_0(Q, \mathbf{u}, \mathbf{u}')[(\mathbf{q} \cdot \mathbf{u})^2 - 1/3]$$
$$R(Q) = (3/2) \int d\mathbf{u} \int d\mathbf{u}' X(Q, \mathbf{u}, \mathbf{u}')[(\mathbf{q} \cdot \mathbf{u})^2 - 1/3]$$
$$T_0(Q) = (9/4) \int d\mathbf{u} \int d\mathbf{u}' X_0(Q, \mathbf{u}, \mathbf{u}')$$
$$\times [(\mathbf{q} \cdot \mathbf{u})^2 - 1/3][(\mathbf{q} \cdot \mathbf{u}')^2 - 1/3]. \tag{C.8}$$

Note that the matrices R and R_0 are not symmetric in the case of copolymers where one of the blocks is flexible and the other one is rigid.

We integrate Eq. (C.3) over u and u' to obtain:

$$X(Q) = X_0(Q) - X_0(Q) \cdot W_0 \cdot X(Q)/k_B T + (2/3)R_0^T(Q) \cdot W_1 \cdot R(Q)/k_B T \tag{C.9}$$

where $R_0^T(Q)$ is the transpose matrix and we have used $[\mathbf{qq} - \mathbf{I}/3]:[\mathbf{qq} - \mathbf{I}/3] = 2/3$. First, multiplying Eq. (C.3) by $[(\mathbf{q} \cdot \mathbf{u})^2 - 1/3]$ and then integrating over u and u' gives another set of equations:

$$R(Q) = R_0(Q) \cdot W_0 \cdot X(Q)/k_B T + (2/3)T_0(Q) \cdot W_1 \cdot R(Q)/k_B T \tag{C.10}$$

These sets of coupled Eqs. (C.9, C.10) can be solved by eliminating $R(Q)$ in order to obtain after a few manipulations:

$$X = \{I + X_0 \cdot W_0 + (2/3)R_0^T \cdot W_1 \cdot M^{-1} \cdot R_0 \cdot W_0\}^{-1}$$
$$\times \{X_0 + (2/3)R_0^T \cdot W_1 \cdot M^{-1} \cdot R_0\} \tag{C.11}$$

where the (Q) argument and the temperature $(k_B T)$ have been dropped for notation convenience and $M = [I - (2/3)T_0 \cdot W_1/k_B T]$ has been defined. Note that the isotropic case (i.e., if orientational correlations were neglected) is obtained when $W_1 = 0$ as:

$$X(Q)^{-1} = X_0(Q)^{-1} + W_0/k_B T \tag{C.12}$$

which is the result for multicomponent compressible blends of flexible polymers.

Appendix D: Incompressible Multicomponent Mixture of Stiff Polymers

Using the matrix notation approach [13–15] that was introduced to describe multicomponent (here also consider n components) flexible polymer systems, the RPA equations are reviewed here for an incompressible stiff polymer mixture. As before, the idea is to isolate a "matrix" component (denoted component M) from the "rest" of the blend (denoted R). The various correlations are described through a scalar part $X_{MM}^0(Q)$, a vector part $X_{MR}^0(Q)$ and a matrix part $X_{RR}^0(Q)$, and similarly for potentials W's. The RPA equations for the n-vector fluctuating densities $\langle \rho(u) \rangle$ are:

$$\langle \rho(u) \rangle = -\int du' X_0(u, u')[U/k_B T + \lambda E$$
$$+ \int du'' W(u', u'') \langle \rho(u'') \rangle / k_B T] \tag{D.1a}$$

$$\langle \rho(u) \rangle = -\int du' X(u, u') U/k_B T \tag{D.1b}$$

where $\langle \rho(u) \rangle = \mathrm{Col}[\langle \rho_R(u) \rangle, \langle \rho_M(u) \rangle]$, E is an n-vector with all terms equal to unity and the Q dependence has been omitted for simplicity in notation. λ is a Lagrange multiplier that is to be determined using the incompressibility constraint:

$$\langle \rho_M \rangle + E_R^T \cdot \langle \rho_R \rangle = 0 . \tag{D.1c}$$

where $\langle \rho_M \rangle = \int du \langle \rho_M(u) \rangle$ and $\langle \rho_R \rangle = \int du \langle \rho_R(u) \rangle$.

Following the same procedure as in the previous appendix, we obtain sets of equations for the orientational moments which are solved to give:

$$\{X_0 + (2/3)R_0^T \cdot W_1 \cdot M^{-1} \cdot R_0\}^{-1} \{I + X_0 \cdot W_0$$
$$+ (2/3)R_0^T \cdot W_1 \cdot M^{-1} \cdot R_0 \cdot W_0\} \langle \rho \rangle = -[U + \lambda E] \tag{D.2}$$

($k_B T$ has been omitted) along with:

$$X^{-1} \langle \rho \rangle = -U \tag{D.3}$$

Here also, we use the $n \times (n-1)$ matrix P to apply the incompressibility constraint and obtain:

$$X_{RR}^{-1} = P^T \cdot \{X_0 + (2/3)R_0^T \cdot W_1 \cdot M^{-1} \cdot R_0\}^{-1}$$
$$\times \{I + X_0 \cdot W_0(2/3)R_0^T \cdot W_1 \cdot M^{-1} \cdot R_0 \cdot W_0\} \cdot P \tag{D.4}$$

This result generalizes the one obtained in Appendix B to include chain stiffness.

Acknowledgements. Many helpful discussions with the following scientists are appreciated: B. Bauer, R. Briber, E. DiMarzio, J. Douglas, C.C. Han, A.I. Nakatani, M. Tombakoglu, D. Waldow and W.L. Wu. Ecouragements from Profs A.Z. Akcasu and H. Benoit are also greatly valued.

11 References

1. Kirste RG, Kruse WA, Schelten J (1973) Makromol Chem 162: 299
2. Flory PJ (1953) Principle of polymer chemistry, Cornell University Press, Ithaca, NY
3. De Gennes PG (1970) J Physique 31: 235; (1980) J Chem Phys 72: 4756
4. De Gennes PG (1979) Scaling concepts in polymer physics, Cornell University Press, NY
5. Zimm B (1946) J Chem Phys 14: 164; (1948) 16: 1093
6. Williams CE, Nierlich M, Cotton JP, Jannink G, Boue F, Daoud M, Farnoux B, Picot C, de Gennes PG, Rinaudo M, Moan M (1979) J Polym Sci, Polym Phys Lett 17: 379
7. Akcasu AZ, Summerfield GC, Jahshan SN, Han CC, Kim CY, Yu H (1980) J Polym Sci, Polym Phys Ed 18: 863
8. King JS, Boyer W, Wignall GD, Ullman R (1985) Macromolecules 18: 709
9. Sanchez IC (1989) J Chem Phys 93: 6983
10. Freed KF (1987) Renormalization group theory of macromolecules, Wiley-Interscience, New York
11. Benoit H, Wu WL, Benmouna M, Mozer B, Bauer B, Lapp A (1985) Macromolecules 18: 986
12. Benoit H, Benmouna M, Wu WL (1990) Macromolecules 23: 1511
13. Akcasu AZ, Tombakoglu M (1990) Macromolecules 23: 607
14. Tombakoglu M (1991) PhD Thesis, U. of Michigan
15. Akcasu AZ (1992) Private Communication (to be published)
16. Ijichu Y, Hashimoto T (1988) Polymer Comm 29: 135. Eq (15) of this paper should not contain the factor of 2
17. Tomalia DA, Baker H, Dewald J, Hall M, Kallos G, Martin S, Roeck J, Ryder J, Smith P (1986) Macromolecules 19: 2466
18. Bauer BJ, Briber RM, Han CC (1989) Macromol 22: 940
19. Benmouna M, Akcasu AZ, Daoud M (1980) Macromolecules 13: 1703
20. Hammouda B, Garcia Molina JJ, Garcia de la Toree JG (1985) J Chem Phys 85: 4120
21. Daoud M (1977) These de Doctorat, University de Paris VI
22. Benoit H, Hadzioannou G (1988) Macromolecules 21: 1449
23. Casassa EF (1965) J Polym Sci, Part A3: 605
24. Benoit H (1953) J Polym Sci 11: 561
25. Hammouda B (1992) J Polym Sci, Polym Phys Ed (in press)
26. Burchard W, Kajiwara K, Nerger D (1982) J Polym Sci, Polym Phys Ed 20: 157
27. Mazur J, McCrackin F (1981) Macromolecules 14: 1214
28. Graessley W (1974) Adv Polym Sci 16: 1
29. Benoit H, Benmouna M (1984) Polymer 25: 1059
30. Shibayama M, Yang H, Stein RS, Han CC (1985) Macromolecules 18: 2179
31. Han CC, Bauer BJ, Clark JC, Muroga Y, Matsushita, Y, Okada M, Tran-Cong Q, Chang T (1988) Polymer 29: 2002
32. Bates FS, Wignall GD (1986) Phys Rev Lett 57: 1429
33. Sakurai S, Hasegawa H, Hashimoto T, Hargis IG, Aggarwal SL, Han CC (1990) Macromolecules 23: 451
34. Trask CA, Roland CM (1989) Macromolecules 22: 256
35. Schweizer KS, Curro JG (1989) J Chem Phys 91: 5059
36. Doi M (1981) J Polym Sci, Polym Phys Ed 19: 229
37. Shimada T, Doi M, Okano K (1988) J Chem Phys 88: 2815
38. Doi M, Shimada T, Okano K (1988) J Chem Phys 88: 4070
39. Leibler L (1980) Macromolecules 13: 1602
40. Holyst R, Schick M (1992) J Chem Phys 96: 721
41. Holyst R, Schick M (1992) J Chem Phys 96: 731. In this paper, freely "jointed" chains are referred to as freely "rotating" chains

42. Hammouda B, Nakatani AI, Waldow DA, Han C (1992) Macromolecules 25: 2903
43. Hammouda B, Briber R, Bauer B (1992) Polym Comm 33: 1785
44. Ciferri A, Krigbaum WR, Meyer RB (eds) (1982) Liquid Crystals, Academic, New York
45. Certain commercial materials and equipment are identified in this paper in order to specify adequately the research procedure. In no case does such identification imply recommendation or endorsement by the National Institute of Standards and Technology, nor does it necessarily imply that the items described are the best available for the purpose.

Editor: H. Benoit
Received January 1992

X-Ray Photoelectron Spectroscopic Studies of Electroactive Polymers

E.T. Kang[1], K.G. Neoh[1] and K.L. Tan[2]
[1] Department of Chemical Engineering, National University of Singapore, Kent Ridge, Singapore 0511
[2] Department of Physics, National University of Singapore, Kent Ridge, Singapore 0511

The electroactive polymers most commonly studied during the past decade include polyacetylene, polyaniline, polypyrrole, polythiophene, polyphenylene, poly(phenylene sulfide), poly(phenylene vinylene) and some non-conjugated polymers, such as the polyvinylpyridine and poly(N-vinyl-carbazole). These polymers along with their analogs and derivatives have been selected to illustrate the type and level of information which can be obtained by the X-ray photoelectron spectroscopic (XPS) technique. It is demonstrated that XPS provides an excellent tool for evaluating the three most important physicochemical properties associated with these polymers, viz., the intrinsic structure, the charge transfer interaction, and the stability and degradation behavior. Some future research directions may involve surface modified or functionalized materials, as well as the application of more surface-sensitive techniques, such as secondary ion mass spectroscopy (SIMS) and scanning tunneling microscopy (STM), which allow us to probe the reactive surfaces and interfaces of the electroactive polymers.

Advances in Polymer Science, Vol. 106
© Springer-Verlag Berlin Heidelberg 1993

Abbreviations

Polymers

$(CH)_x$	polyacetylene
PPA	polyphenylacetylene
Poly(o-Me$_3$SiPA)	poly[[o-(trimethylsilyl)phenyl]acetylene]
PPY	polypyrrole
DP-PPY	deprotonated (25%) polypyrrole
PPY°	fully reduced polypyrrole
PAN	polyaniline
PNA	pernigraniline
EM	emeraldine
LM	leucoemeraldine
NA	nigraniline
PTH	polythiophene
PPS	poly(phenylene sulfide)
PBT	polybenzothiophene
PBiT	poly(2,2′-bithiophene)
PMeT	poly(3-methylthiophene)
PHeT	poly(3-hexylthiophene)
PPV	poly(phenylene vinylene)
PPP	poly(p-phenylene)
P2VP	poly(2-vinylpyridine)
P4VP	poly(4-vinylpyridine)
PVK	poly(N-vinylcarbazole)

Experimental Techniques

XPS	X-ray photoelectron spectroscopy
ESCA	electron spectroscopy for chemical analysis
UPS	ultra-violet photoelectron spectroscopy
SIMS	secondary ion mass spectroscopy
STM	scanning tunneling microscopy
IR	infra-red
UV	ultra-violet

Miscellanea

BE	binding energy
KE	kinetic energy
CT	charge transfer
DDQ	2,3-dichloro-5,6-dicyano-p-benzoquinone
TCNE	tetracyanoethylene

TCNQ	7,7,8,8-tetracyano-*p*-benzoquinone
o-chloranil	tetrachloro-*o*-benzoquinone
p-chloranil	tetrachloro-*p*-benzoquinone
o-bromanil	tetrabromo-*o*-benzoquinone
p-fluoranil	tetrafluoro-*p*-benzoquinone
NMP	*N*-methylpyrrolidinone

1 Introduction

The synthesis and characterization of electroactive polymers have become one of the most important areas of research in polymer science during the past one and a half decades [1]. The most commonly studied electroactive polymers include polyacetylene $(CH)_x$, polyaniline (PAN), polypyrrole (PPY), polythiophene (PTH), poly(phenylene sulfide) (PPS), poly(phenylene vinylene) (PPV) and their derivatives and analogs. Some non-conjugated electroactive polymers, such as the vinylpyridine and vinylcarbazole polymers, have also been actively studied. In the investigation of a typical electroactive polymer system, one endeavors for a better understanding of the three most important physicochemical properties, viz., (a) the intrinsic structure, (b) the charge transfer (CT) interaction, and (c) the stability and degradation behavior of the polymer system. It is the purpose of this article to demonstrate, by examples and case studies, that these basic properties can be advantageously investigated by X-ray photoelectron spectroscopy (XPS), an analytical technique also known as electron spectroscopy for chemical analysis (ESCA).

With the recognition of its usefulness as an analytical tool, XPS has been extensively used in the characterization of electroactive polymers and their complexes. However, a survey of the literature on XPS studies of electroactive polymers reveals a surprisingly large number of discrepancies among the reported data. The problem is aggravated by the fact that a substantial number of published XPS results suffers from the common pitfall of casual analysis of the spectral data. Thus, much of the important information is either lost or remains untapped in many cases. Undoubtedly, peak synthesis or deconvolution of the XPS core-level spectra largely remains a personal judgement, but with careful energy referencing and calibration, as well as careful comparison with model compounds, it is possible in many cases to resolve various species and components contained in a core-level spectrum unambiguously. In addition, a large amount of core-level energy information has been accumulated over the years [2].

Salaneck [3] presented in the mid-1980s a fine review on the application of photoelectron spectroscopic techniques to the study of electroactive polymers. However, a substantial number of new and significant XPS studies have since appeared, in conjunction with the ever increasing research on new families of electroactive polymers. This is particularly true for aniline polymers and some polyheterocycles. Thus, updating XPS work on electroactive polymers appears to be appropriate. This review will focus mainly on XPS core-level spectra, with some references made to valence band spectra. First, a brief description of the basic principles of XPS is presented for readers who are less familiar with this technique. Next, the type and level of information that its application provides for the elucidation of the intrinsic structure, the CT interaction, and the stability and degradation behavior of each family of electroactive polymers are presented and discussed in detail. Finally, some future directions for the applications of

XPS and other surface analysis techniques to the study of electroactive polymers and their molecular modifications are pointed out. We assume that the readers are familiar with the historical development of electroactive polymers since the pioneering work on $(CH)_x$ [4, 5].

2 Principles of X-Ray Photoelectron Spectroscopy (XPS)

The basic principles [6] of the XPS technique are depicted in Fig. 1. An incident X-ray photon (hv) causes an electron to be ejected from one of the core electronic levels. The photoemitted core-electron escapes with a certain kinetic energy (KE), which is governed by the binding energy (BE) of the core-electron in the sample and the work function of the spectrometer (φ) according to

$$KE = hv - BE - \phi$$

where the zero BE is taken to be the Fermi level (E_F) of the sample. When a core-hole is produced, it is either filled by an outer electron with the accompanied emission of a photon (X-ray) or by the emission of a secondary electron (Auger electron) through a radiationless transition.

The Auger and photoemitted electrons of interest have relatively low KE (50 to 2000 eV) and have a high probability of undergoing inelastic collision with an atom in the matrix. Thus, only those photoelectrons generated near the sample surface (< 10 nm) will contribute to XPS signals. Detailed experimental values of the inelastic mean free path for electrons having the energies important in XPS have been reported [7]. Furthermore, detailed descriptions of the hardware and instrumentation involved in the XPS technique are also available [8].

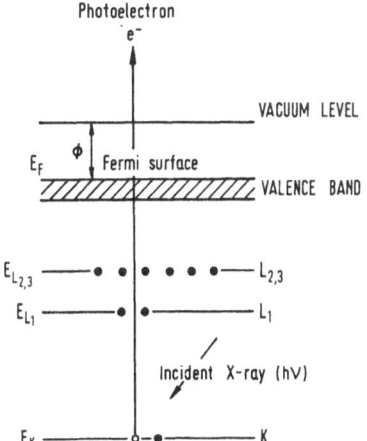

Fig. 1. Energy-level diagram showing the electron transition process in XPS

2.1 Core-Level Spectra

Siegbahn et al. [9] were the first to demonstrate the applicability of the photoelectron spectroscopic techniques to chemical structure analysis. The pioneering work of Clark [10], Dilks [11] and Briggs [12] on the application of the XPS technique to polymers, coupled with non-empirical molecular orbital calculations, established the indisputable capabilities of the technique to elucidate many important aspects of polymer-surface chemistry. The level of information that can be derived from the principal features of the typical XPS core-level spectra of polymers is summarized in Table 1. Thus, from the characteristic BEs of the photoelectrons, the elements involved can be identified. The peak intensity can be directly related to the atomic concentration in the sample. One of the most important features of XPS is its ability to measure shifts in the BE of core-electrons accompanying changes in the chemical environment. For example, when an electron is withdrawn from the atom (oxidation), BE increases and vice versa. Furthermore, a sudden change in the effective nuclear charge due to the loss of a shielding electron during the photoionization process may cause one of the valence electrons to be excited to an unfilled higher energy level ($\pi \rightarrow \pi^*$ transition). This transition is accompanied by a quantum loss in the kinetic energy of the photoelectron and will result in a discrete satellite photoelectron line at a higher BE than the primary line (the so-called 'shake-up' phenomenon). Thus, the shake-up satellite structures provide additional information about unsaturated polymers. The four main spectral features of XPS are best illustrated in the C $1s$ core-level spectrum of poly(ethylene terephthalate) film shown in Fig. 2 [13]. The overall spectrum (solid line) is resolved (deconvoluted) into four Gaussian component peaks (dashed lines). The main peak at 285.0 eV is assigned to the six carbon atoms of the benzene ring, the peak at 286.8 eV to the two ester carbons, and the third peak at 289.0 eV to the two carboxyl carbons. The 6:2:2 ratio of carbon functionality is reflected approximately by the component peak area ratios. A low-intensity shake-up satellite arising from the aromatic ring at about 291.5 eV is also discernible.

Table 1. Information associated with the main features of the XPS spectra of polymers

Spectral feature	Information
Core-level Binding Energy (Peak Position)	Chemical Species Identification
Binding Energy Shift (Chemical Shift)	Redox State, Chemical Environment
Peak-area Ratio	Stoichiometry
Shake-up Satellite	$\pi \rightarrow \pi^*$ Transition

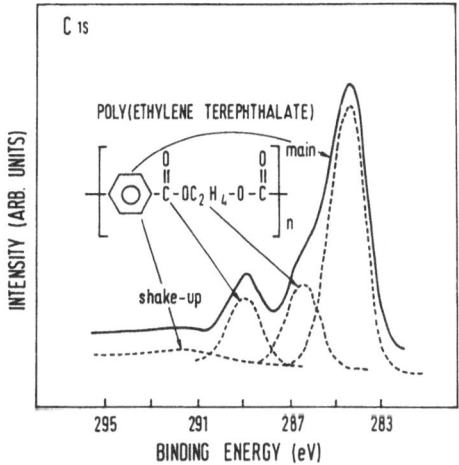

Fig. 2. C1s core-level spectrum of a poly(ethylene terephthalate) film

2.2 Valence Band Spectra

The valence band region is often loosely defined as the first 20 eV of the BE scale, involving electrons in delocalized or bonding orbitals [14]. If the KE of the photoelectron from the occupied state in this region falls within the structured unoccupied states, the observed intensity will then reflect the transition probability, determined by the empty and filled states. This is the basis of the ultra-violet photoelectron spectroscopy (UPS), in which the valence-level spectrum shows a strong dependence on photon energy. The XPS valence band spectra, on the other hand, relate closely to the initial occupied states, because the KE of the valence photoelectrons is such that the final states are usually quite devoid of structure. In the case of polymers, the XPS valence band spectra provide a finger-printing capacity.

The XPS valence band spectrum of an electroactive polymer/dopant complex can be understood in terms of the known valence band spectra of the monomer and the dopant. Thus, the valence band spectroscopy confirms the core-level results by identifying the monomer and dopant species [15].

3 XPS of Electroactive Polymers

3.1 Intrinsic Structures

3.1.1 Acetylene Polymers

The recent interest in electroactive polymers started with the preparation of polyacetylene $(CH)_x$ film [4] and the discovery of the exceptional electrical properties of doped film [5]. Most XPS core-level studies of acetylene polymers have been devoted to the polymer CT complexes. (See Sect. 3.2.1 below). As a main member of the family, pristine $(CH)_x$ contains only carbon atoms in one single chemical environment. Thus, the carbon core-level spectrum of pure $(CH)_x$ is expected to provide only a very limited amount of information about the intrinsic structure of the polymer. Furthermore, the valence band spectrum of pure $(CH)_x$ shows no distinct XPS signal, because the $C2p$ derived molecular orbitals in the valence band BE region have very small cross sections at X-ray photon energies [16]. Nevertheless, the XPS core-level spectrum of $(CH)_x$ gives some valuable information about the structural modifications of the film surface, which result from interactions with the environment. Figure 3 shows the $C1s$ core-level spectrum of an as-prepared $(CH)_x$ film [17]. Thus, owing to the reactive nature of the polymer surface, most of the $(CH)_x$ film surfaces are oxidized to a small extent. The formation of covalent bonds between carbon and oxygen reduces the effective conjugation length of the polymer and renders it less conductive. As will be shown in Sect. 4, the intrinsic structure of $(CH)_x$ below the oxidized film surface can be partially unveiled if the XPS technique is supplemented by other surface analysis techniques having depth profiling capability. Similar structural modifications at the polymer surface have also been observed in the XPS studies of many substituted acetylene polymers, such as the photoconductive polyphenylacetylene (PPA) [18], although the substituted acetylene polymers generally exhibit a much better environmental stability than $(CH)_x$ [19, 20].

3.1.2 N-Containing Polymers

Two nitrogen-containing electroactive polymers, polypyrrole (PPY) [21] and polyaniline (PAN) [22], have been of particular interest because of their environmental stability, high electrical conductivity and interesting redox properties associated with the chain heteroatoms. More importantly, PAN has been found to exhibit solution processability [23, 24] and partial crystallinity [25, 26].

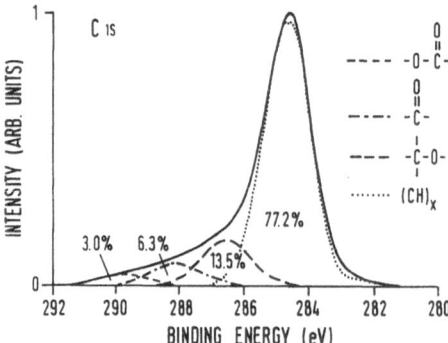

Fig. 3. $C1s$ core-level spectrum of an as-prepared $(CH)_x$ film

The aniline polymers have the general formula $[(-B-NH-B-NH-)_y$ $(-B-N=Q=N-)_{1-y}]_x$, in which B and Q denote the C_6H_4 rings in the benzenoid and quinonoid forms, respectively [22]. Thus, they are basically poly(p-phenyleneimineamine)s, in which the intrinsic oxidation states range from the fully oxidized pernigraniline (PNA, $y = 0$) to the fully reduced leucoemeraldine (LM, $y = 1$). The 50% oxidized polymer has been termed emeraldine (EM, $y = 0.5$). PAN can achieve its highly conductive state either through the protonation of the imine nitrogens (=N– structure) in its EM oxidation state or through the oxidation of the amine nitrogens (–NH– structure) in the fully reduced LM. The synthesis and the physicochemical properties of PAN have recently been reviewed [27].

XPS plays a role as a truly unique tool in quantitative evaluation of the various intrinsic redox states of PAN. Earlier XPS studies on electrochemically prepared PAN and its substituted derivatives [28, 29] compared the N$1s$ photoelectron peak width (full width at half maximum or FWHM) for the protonated PAN with that for the EM base. The increase in the N$1s$ linewidth of the EM base was associated with the presence of chemically different nitrogen atoms. A number of more recent studies [30, 31] on chemically synthesized PAN have demonstrated that the proportions of quinonoid imine (=N– structure), benzenoid amine (–NH– structure) and positively charged nitrogens corresponding to a particular oxidation state and protonation level of PAN can be quantitatively differentiated in the properly curve-fitted N$1s$ core-level spectrum. They correspond to peak components with BEs at about 398.2 $\pm$ 0.1 eV, 399.4 $\pm$ 0.1 eV and > 400 eV. Figure 4(a) to Fig. 4(c) show, respectively, the N$1s$ core-level spectra for a nigraniline (NA, 75% intrinsically oxidized), an EM base and an LM [32]. The consistency in the BE assignments and peak synthesis is readily evidenced by the presence of about equal proportions of the imine and amine N$1s$ components in the EM base and only a single nitrogen environment at about 399.4 eV in the fully reduced LM. These BE assignments were confirmed by at least two later independent studies on electrochemically synthesized PAN [33, 34]. The residual high BE tail above 400 eV in the neutral NA and EM base may have resulted at least in part from surface oxidation products. Although methods for the preparation of neutral and fully oxidized PNA have been proposed [35], no N$1s$ core-level spectrum of PNA has yet been reported. For electrochemically oxidized PAN, the maximum intrinsic oxidation state, found on the curve-fitted N$1s$ spectrum, is also only of the order of 75%, corresponding to the NA oxidation state [33]. Effects of chemical synthesis conditions on the intrinsic structure of PAN have recently been compared [36].

Figure 5(a) and Fig. 5(b) show, respectively, the N$1s$ and S$2p$ core-level spectra of an EM base after protonation by $1 \, \text{mol} \, l^{-1} \, H_2SO_4$ [37]. The S/N ratio of the resulting complex is 0.52. Thus, all the imine nitrogens have been transformed into the positively charged nitrogens, conforming to the general belief that protonation occurs preferentially at the imine repeating units [22]. Furthermore, a close balance is observed between the proportion of positively

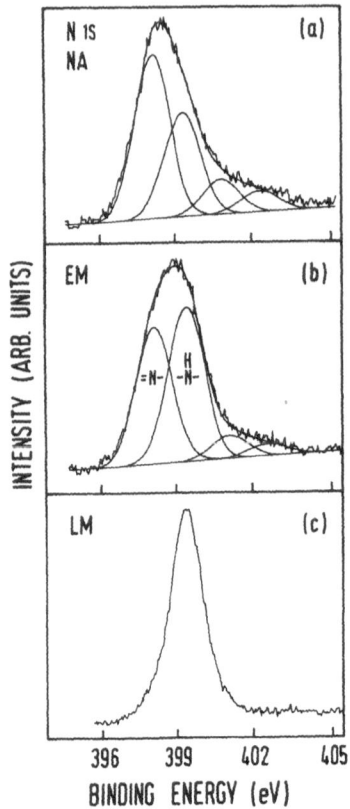

Fig. 4a–c. N$1s$ core-level spectra of (**a**) NA, (**b**) EM base, and (**c**) LM

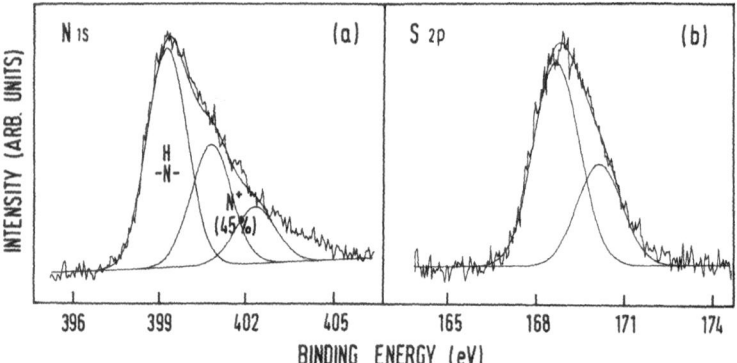

Fig. 5a, b. Curve-fitted (**a**) N$1s$ and (**b**) S$2p$ core-level spectra for H$_2$SO$_4$ protonated EM with an S/N ratio of 0.52

charged nitrogens and the S/N ratio of the complex. Charge neutrality consid-
erations suggest that the sulfate anion must exist predominantly as the mono-
valent HSO$_4^-$ species.

Effects of substitution on the intrinsic redox state of PAN have also been investigated, with the results as summarized in Table 2 [38]. Thus, the types of the halogen substituent, the positions of the substitution and the protonic acid media do not significantly change the intrinsic oxidation state of the polymer. However, the intrinsic oxidation state of the base polymers, as suggested by the imine:amine ratio, is only one half of that of the unsubstituted EM base. The somewhat lower degree of protonation by HCl than by H_2SO_4 is probably attributable to the volatile nature of the former in the ultra-high vacuum environment during XPS measurements. Furthermore, no halogen elimination took place in all of the chemically prepared polymer complexes, contrary to the observations for electrochemically prepared complexes [29].

For completeness, we note a recent XPS study [39] on PAN which found that the N/s BEs for the imine and amine nitrogens differed by only 0.1 eV and therefore failed to be resolved. Yet another study [40] assigned the N1s peak component at 398.2 and 399.2 eV to neutral amine and imine moieties, respectively, differing from the assignments shown in Fig. 4. The potential model calculations in this study also concluded that the bulk of positive charges on the PAN backbone reside on the amine nitrogens. This conclusion again contradicts the well-established fact that protonation occurs preferentially at the imine units [22]. Although the imine and amine nitrogen peak assignments were corrected in a later study [41], the reported high level of protonated amine units remains to be justified.

Although the intrinsic redox states of polypyrrole (PPY) are much less known than those of PAN, Inganäs et al. [42] first revealed that base treatment of oxidized PPY complexes led to a deprotonation process. This is readily seen by the loss of the N1s high BE tail above 400.5 eV and the appearance of a low BE shoulder shifted by about -2 eV from the main pyrrolylium nitrogen (–NH– structure) peak at about 399.7 eV. The low BE component at about 397.7 eV was attributed to the deprotonated pyrrolylium nitrogens or the imine-like nitrogens ($=N-$ structure). The presence of these deprotonated pyrrolylium nitrogens has also been reported for PPY synthesized either chemically [43] or electrochemically [44, 45] under a less oxidative environment, or during the interaction of PPY with ammonia [46].

Figure 6 compares the intrinsic redox state in PPY with that in PAN [47]. Figure 6(a) shows the N1s core-level spectrum for a PPY/chloride complex with a Cl/N ratio of 0.33, synthesized from $FeCl_3 \cdot 6H_2O$. An N1s core-level spectrum similar in line-shape but with a more pronounced high BE component is observed for the acid-protonated EM oxidation state of PAN, such as the emeraldine hydrochloride (EM/HCl) (Fig. 6(b)). In the case of the EM/HCl complex, a Cl/N ratio of about 0.5 is observed when all the imine units are preferentially protonated. Figure 6(c) and Fig. 6(d) show, respectively, the N1s core-level spectra for the PPY/chloride and EM/HCl complexes after 'undoping' by 0.5 mol l^{-1} NaOH. The deprotonation process in each complex is consistent with the appearance of a low BE N1s component, attributable to the imine nitrogens. In the case of deprotonated PPY (DP-PPY), the imine nitrogens

Table 2. XPS results and stoichiometries of various halogen-substituted polyaniline complexes

Monomer	Polymerization medium	XPS surface stoichiometry				Proportion of −NH−	N^+	Conductivity (S/cm)
		Cl/N	Br/N or I/N	S/N	$=N^{-}$[b]			
2-chloroaniline	H_2SO_4	1.05 (1.09)[a]	–	0.29	0.13 (0.30)	0.58 (0.62)	0.29 (0.08)	10^{-6}
3-chloroaniline	H_2SO_4	1.13 (1.15)	–	0.17	0.13 (0.24)	0.72 (0.70)	0.15 (0.06)	10^{-7}
4-chloroaniline	H_2SO_4	1.11 (1.06)	–	0.32	0.0 (0.31)	0.65 (0.69)	0.35 (0.0)	10^{-7}
2-chloroaniline	HCl	1.57 (1.45)	–	–	0.28 (0.30)	0.61 (0.62)	0.11 (0.08)	$< 10^{-10}$
3-chloroaniline	HCl	2.68 (2.50)	–	–	0.19 (0.24)	0.68 (0.68)	0.13 (0.08)	$< 10^{-10}$
4-chloroaniline	HCl	2.07 (1.96)	–	–	0.18 (0.28)	0.71 (0.64)	0.11 (0.08)	$< 10^{-10}$
2-bromoaniline	H_2SO_4	–	1.04 (1.03)	0.30	0.11 (0.26)	0.64 (0.67)	0.25 (0.07)	10^{-7}
3-bromoaniline	H_2SO_4	–	1.10 (1.20)	0.24	0.12 (0.28)	0.64 (0.64)	0.24 (0.08)	10^{-8}
4-bromoaniline	H_2SO_4	–	0.93 (1.04)	0.45	0.0 (0.28)	0.62 (0.64)	0.38 (0.08)	10^{-8}
2-bromoaniline	HCl	0.32 (0.26)	1.05 (1.07)	–	0.27 (0.27)	0.63 (0.66)	0.10 (0.07)	$< 10^{-10}$
2-iodoaniline	H_2SO_4	–	1.10 (1.12)	0.22	0.17 (0.33)	0.56 (0.57)	0.27 (0.10)	10^{-8}
2-iodoaniline	HCl	0.40 (0.30)	1.09 (0.95)	–	0.26 (0.30)	0.66 (0.63)	0.08 (0.07)	$< 10^{-10}$
4-iodoaniline	HCl	0.62 (0.49)	1.11 (1.0)	–	0.23 (0.28)	0.66 (0.64)	0.11 (0.08)	$< 10^{-10}$

[a] Number in ( ) indicates the value for the sample deprotonated by 0.5 mol l^{-1} NaOH.
[b] Neutral imine units.

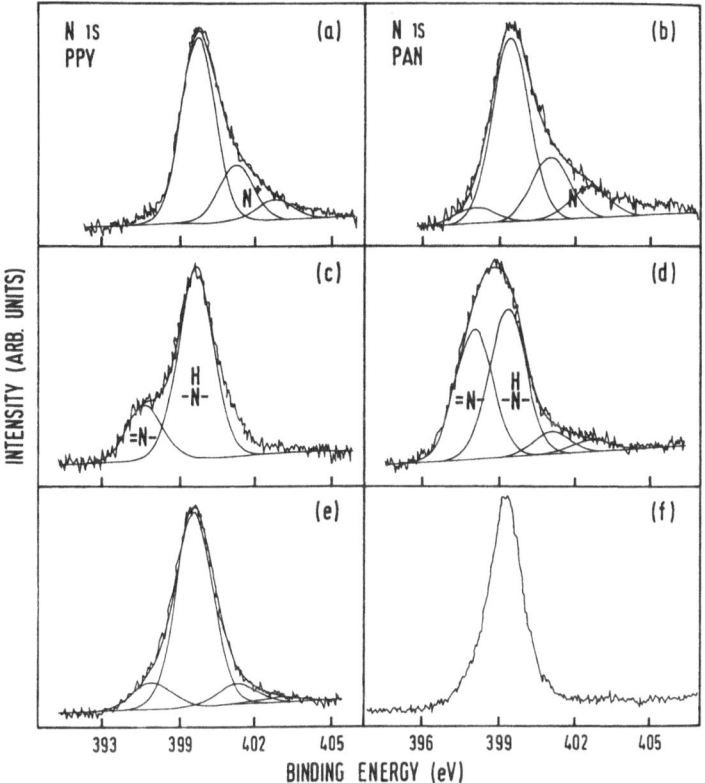

Fig. 6a–f. N$1s$ core-level spectra of (**a**) PPY/chloride (Cl/N = 0.33); (**b**) EM/HCl (Cl/N = 0.44); (**c**) DP-PPY; (**d**) EM base; (**e**) PPY°; (**f**) LM

account for about 25% of the total nitrogens. About 20–25% of the imine-like structure is generally observed for other deprotonated PPY/anion complexes, such as the PPY/iodide and PPY/perchlorate complexes [43]. Thus, the structure of DP-PPY consists of about 25% oxidized neutral imine-like units. This percentage is only one half of that of its EM base counterpart, but is consistent with the fact that the 'doping' level of the PPY complexes is usually only one half of that of the EM salt.

Reprotonation of the DP-PPY and EM base by 1 mol l^{-1} HCl brings the polymers back to their positively charged and highly conductive state, with the respective N$1s$ core-level spectra similar to those shown in Fig. 6(a) and Fig. 6(b). The EM oxidation state of PAN can also be completely transformed to the fully reduced LM upon treatment with phenylhydrazine (Fig. 6(f)). Similar treatment of DP-PPY also gives rise to a fully reduced polymer or PPY°, as evidenced from the disappearance of the low BE shoulder in the N$1s$ core-level spectrum of Fig. 6(e). The reduction process, however, is more sluggish in this case. The PPY° in turn is susceptible to partial oxidation by electron acceptors, such as the halogens [48]. Treatment of this re-oxidized PPY with a base again

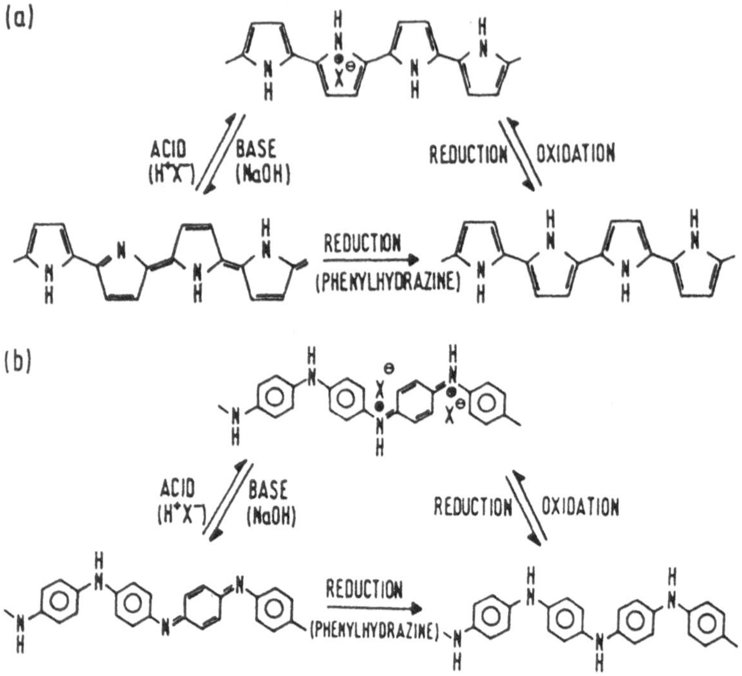

(a)

(b)

Fig. 7a, b. Interconversions between various redox states in (a) PPY and (b) PAN

invites a deprotonation process. This may be compared to the observation that oxidation of the fully reduced amine units in LM by halogens gives rise to a structure equivalent to that of the acid-protonated imine units in EM [22, 48].

In summary, XPS data allow us to conclude that proton modification of nitrogens in PPY gives rise to a number of intrinsic redox states analogous to those observed in PAN. The behavior of the corresponding redox states in these polymers towards oxidation/reduction and deprotonation/reprotonation are grossly similar. It should be noted at this stage, however, that their nitrogens differ in thermal degradation behavior (see Sect. 3.3.2 below). The interconversions between the various redox states associated with the chain nitrogens in PPY and PAN are summarized in Fig. 7 [47].

3.1.3 S-Containing Polymers

Thiophene polymers, in particular, alkyl-substituted polythiophenes (PTH), are some of the conducting polymers being most actively investigated at present. This fact is attributable to their high degree of processability, environmental stability [49, 50] and, in some cases, ability to exhibit reversible electrochromism [51] and thermochromism [52]. Another important family of sulfur-containing polymers is based on poly(phenylene sulfide) (PPS) [53] and its derivatives, such as polybenzothiophene (PBT) obtained by exposure of PPS to AsF$_5$ [54].

A comprehensive study of the core and valence electronic structures in many sulfur-containing electroactive polymers has been reported [55]. The valence band spectra for some of them can be well resolved, as in Fig. 8 which shows the XPS valence band spectra of PPS and PBT. The peaks A, B, C and D in the BE region between 10 and 20 eV serve as the 'fingerprint' of the benzenic backbone and arise mainly from C$2s$ contributions. The peak E contains σ (C$2p$-S$3p$-H$1s$) and the lowest π (C$2p$-S$3p$) bands. The F peak (the highest occupied band) that is more intense in PPS than in PBT is attributable to a larger S$3p$ cross-section of the former in this region. The similarity of the satellite-to-main peak distance between the C$1s$ and S$2p$ core-level spectra suggests that these spectral features reflects the same valence excitation and thus confirm the contribution of the sulfur atom to the highest occupied band.

XPS studies of other S-containing heterocycles, such as PTH, poly-thieno(2,3-b)thiophene (PTH23TH) and polydithienothiophene (PDTT), [55] also indicated that, in the oxidized state, the C$1s$ and S$2p$ core-level spectra appear at a higher BE than the neutral polymer and are characteristic of positively polarized atoms. This result appears to be inconsistent with the fact that, in the oxidized polymer, only one out of every 3 to 4 monomer units is associated with a dopant anion and the C$1s$ and S$2p$ core-level spectra should reveal the simultaneous presence of neutral and charged (or polarized) species (see also Sect. 3.2.1 below).

The electronic structures of PTH, poly(2,2'-bithiophene) (PBiT) and poly-(3-alkylthiophene)s have also been the subject of a number of XPS and UPS

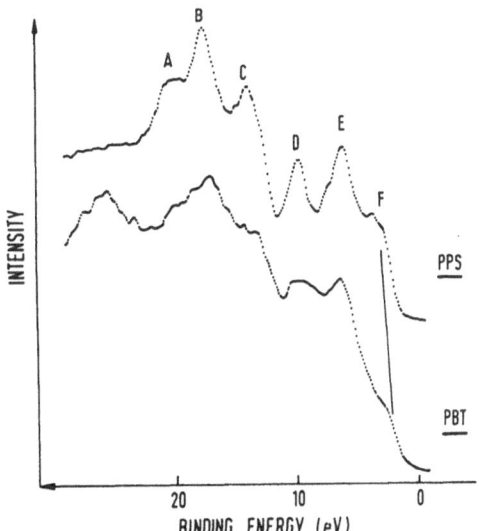

Fig. 8. XPS valence band spectra of PPS and PBT

studies [56, 57]. A systematic evolution of the π bonding orbitals was observed by going successively from thiophene to bithiophene and terthiophene. It finally led to the formation of an intrinsic π bonding band for undoped PTH. A comparison of the XPS gas phase shake-up data with the corresponding data on the solid polymers indicated that with increasing temperature, the electronic structure exhibits increasing oligomeric-like electronic localization effects. This temperature-dependent localization of the π-electron is the electronic basis for the observed thermochromism in poly(3-alkylthiophene)s.

3.1.4 Other Conjugated Polymers

A variety of arylenevinylene polymers, in particular poly(phenylene vinylene) (PPV) and its derivatives, were prepared and shown to become highly conductive when doped [58, 59]. The C$1s$ core-level spectrum of undoped PPV was asymmetric, and was resolved into two components with BEs at about 284.6 eV (80%) and 285.5 eV (20%) [60]. Theoretical study of a low molecular weight analog of PPV, p,p'-$trans$-vinylene-bis(stilbene), suggested that the presence of a non-uniform electron distribution in the polymer chain can give rise to a number of non-equivalent carbon centers. Thus, the C$1s$ components observed in PPV correspond to a non-uniform electron population of carbon atoms within the conjugated electronic system of the polymer. The XPS valence band spectrum of PPV consists of several relatively narrow photoionization peaks beginning at 3 eV. These peaks have been explained in terms of the states predicted from a tight binding model for conjugated chains.

Although the electrical properties of conducting polymers have been extensively studied, little attention has been paid to the structure of the interfacial contact between the polymer and the metallic electrode. Chemical reactions may take place at the interface owing to the reactive nature of most electroactive polymers, as is best demonstrated in the XPS study of the Al/poly(p-phenylene) (PPP) interface [61]. The technique used should be readily applicable to the study of other polymer-metal interfaces. It involves the sputtering of an evaporated Al layer by Ar$^+$ bombardment with controlled flux. Core-level spectra of C, Al and O are recorded and the process is repeated until the C$1s$ spectrum of pure PPP film is obtained. Figure 9 shows the C$1s$ core-level spectrum of PPP and the changes in the C$1s$, O$1s$ and Al$2p$ core-level spectra at the Al/PPP interface upon successive removal of the Al contact. The interfacial layer is defined as the region in which the C$1s$ and Al$2p$ signals appear simultaneously. Comparison of the Al$2p$ line in the Al/PPP interface with that in clean Al metal, which has a BE of about 72.8 eV, suggests that Al is in an oxidized state. The shift of the C$1s$ line at the interface towards lower BE has been attributed to the formation of metal carbides. Furthermore, the O$1s$ lines are consistent with the presence of an Al$_x$O$_y$ structure. Thus, it may be concluded that an Al oxide-carbide complex exists at the Al/PPP interface.

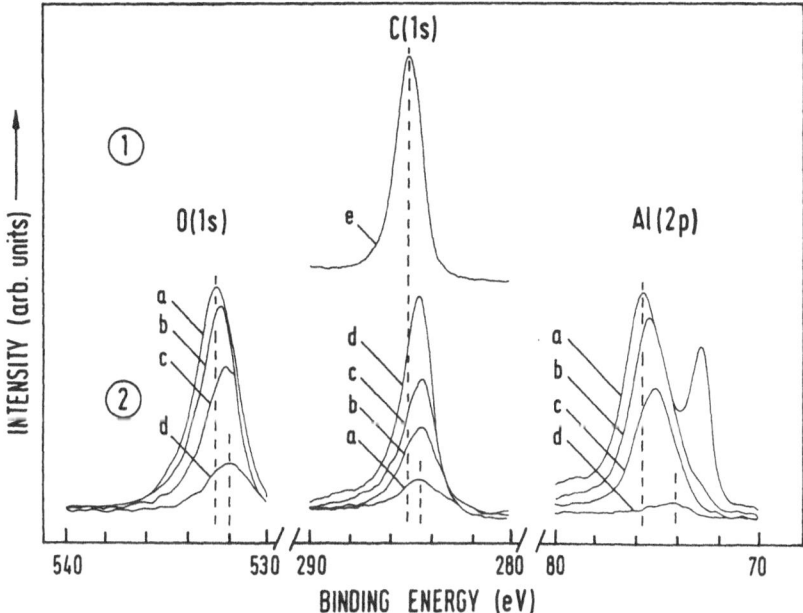

Fig. 9. Changes in the C*1s*, O*1s* and A1 *2p* core-level spectra at the A1/PPP interface upon successive removal of the metal contact by sputtering: (**1**) without A1 layer (curve *e*), (**2**) in the interfacial PPP/A1 layer (curves *a,b,c,d*) with successive removal of the A1 layer

3.2 Charge Transfer Interactions

3.2.1 Conjugated Polymers

XPS provides a powerful and unique tool for the study of CT interactions in organic molecules, as is best illustrated by the earlier work of Ng and Hercules [62] on some chloranil-donor adducts. Most of the earlier XPS work on electroactive organic polymers dealt with the chemical states of inorganic molecular dopants and their CT interactions with the polymers.

3.2.1.1 Acetylene Polymers

The earliest XPS work on modern electroactive polymers appears to be that of Hsu et al. [63] on the chemical states of the dopant in iodine-doped $(CH)_x$ films. In this work, combined XPS and Raman scattering studies revealed the presence of I_3^- and I_5^- species. The latter species resulted from the equilibrium process of the type: $I_2 + I_3^- = I_5^-$. Similar findings were also made in at least two separate studies [64, 16]. The presence of polyiodide anion species was also observed in complexes involving substituted polyacetylenes, such as PPA and poly(3,3-dimethyl-1-butyne) [65].

In the case of bromine-doped *cis*- and *trans*-$(CH)_x$, the positively charged carbon atoms in the metallic domain and the carbon atoms in the undoped low conductivity domain can be distinguished in the $C1s$ core-level spectrum [66]. XPS results have also indicated that the bromine dopant exists as the Br_3^- anion. The dopant is more concentrated in the surface region than in the bulk at low bromine content, but is distributed more homogeneously in the heavily doped film. However, the $Br3d$ core-level spectrum reveals the presence of a considerable amount of covalently bonded bromine in the heavily doped film, arising from the addition of bromine to the C=C double bonds. This addition causes the breakdown of π-electron conjugation and an abrupt drop in electrical conductivity. A similar phenomenon was also observed in the XPS studies of $FeCl_3$-doped $(CH)_x$ [67]. In this case, the dopant existed as the $FeCl_4^-$ species, and a substantial amount of covalently bonded chlorine was detected in the highly doped polymer. Many XPS studies have also been made on AsF_5-doped $(CH)_x$ films [68–70], but the reported data are less consistent (see also Sect. 3.3.2). This result has been attributed to the difference in sample preparation and/or handling [3].

XPS studies have revealed unusual and interesting CT interactions in WCl_6 and $MoCl_5$-doped $(CH)_x$ [71]. A lowering of BE was observed for C, W, Mo and Cl after doping, and it was attributed to an increase in the outershell electron densities of these elements in the complexes. The increase has been explained in terms of the transfer of $p\pi$-electrons of the olefins to the metal d-orbital, followed by the back-donation to the $p\pi^*$ orbital of the olefins and the Cl^- ligands.

Relatively few studies have been made on n-doped $(CH)_x$ [72]. For Na-doped $(CH)_x$, the core-level spectra suggest CT from Na to $(CH)_x$, which leads to the formation of negatively charged carbon, as seen from the appearance of a low BE shoulder in the $C1s$ core-level spectrum. Furthermore, the observed stoichiometric balances indicate that the transferred electrons can spread over several carbon units of the $(CH)_x$ chain.

More recently, the interest of XPS studies has been directed to the CT interactions between environmentally stable substituted polyacetylenes and various inorganic and organic electron acceptors. Although most of the substituted acetylene polymers are poor dark conductors even after doping, some of them possess good photoconductivity [73–75] and high gas permeability [20]. A fairly strong CT interaction has been reported for the $FeCl_3$-doped PPA in tetrahydrofuran [76]. Distinct chemical shifts and signal broadening caused by CT with the dopant have been observed in the $C1s$ core-level spectrum of the complex. Evaluation of the Fe/Cl atomic ratio suggests that the dopant exists as the $FeCl_4^-$ species. However, for the CT interaction in wet acetone, a second chlorine species with a BE of about 200 eV was found, and it was attributed to the presence of covalently bonded Cl species. On the other hand, relatively weak CT interactions between PPA and various organic acceptors were indicated by the lack of distinctive chemical shifts in the core-level BE of the polymer and the dopants [18]. The acceptors used include 2,3-dichloro-5,6-dicyano-*p*-benzoquinone (DDQ), tetracyanoethylene (TCNE), 7,7,8,8-tetracyano-*p*-benzoquinone (TCNQ) and various halobenzoquinones, such as *o*-chloranil, *p*-chloranil, *o*-bromanil and *p*-fluoranil. In the case of Si-containing poly-[[*o*-trimethylsilyl)phenyl]acetylene] [Poly(*o*-Me$_3$SiPA)], the appearance of distinctive partially charged or polarized species in the acceptor [19] suggests an increase in the extent of CT to take place, despite an increase in the bulkiness of the substituent on the polymer chain.

3.2.1.2 The N-Containing Polymers

The presence of heteroatoms in the polymer chains and their direct participation in the CT processes have made the core-level spectroscopic studies of the PPY and PAN complexes particularly fruitful. Serious discrepancies, however, can be found in the earlier XPS studies on PPY complexes. Salaneck et al. [77] reported a single-component N1s spectrum, although it was displaced to a slightly higher BE for the PPY/BF$_4$ complex. On the other hand, Pfluger and Street [15] and Skotheim et al. [44] found N1s spectra with a high BE shoulder for PPY/ClO$_4$ and PPY/BF$_4$ complexes. However, all these studies suggest delocalized positive charges or partially charged nitrogens. Munro et al. [78, 79] made the first attempt to quantitatively correlate the proportion of the N1s high BE component with the anion doping level. For the more fully doped samples, which are less susceptible to surface oxidation and contamination, good correlation can be obtained if it is assumed that all oxidized nitrogen atoms bear a unit positive charge. The presence of unit positive charges on the pyrrolylium nitrogens has also been favored in many subsequent studies involving oxidized PPY complexes prepared chemically in the presence of halogens [80], Fe(III) salts [48] and organic electron acceptors [81, 82]. All these studies revealed the coexistence of neutral and positively charged nitrogens in the N1s core-level spectrum, along with a good correlation between the proportion of positively charged nitrogen and the anion/N ratio.

Table 3 shows the effect of progressive NaOH treatment on a PPY/chloride complex synthesized in the presence of FeCl$_3$ · 6H$_2$O in ethanol. The N1s and

Table 3. Effect of progressive NaOH treatments on a PPY complex synthesized from FeCl$_3$ · 6H$_2$O in ethanol as revealed by XPS

Sample	Total Cl/N (XPS)	Chemical States[a]			Proportions[b]			Conductivity (S/cm)
		Cl$^-$/N	Cl*/N	–Cl/N	=N–	–NH–	N$^+$	
1 (Untreated)	0.41	0.21	0.14	0.06	0.03	0.68	0.29	38
2	0.24	0.17	0.04	0.03	0.08	0.68	0.23	3
3	0.15	0.10	0.03	0.02	0.14	0.68	0.17	0.1
4	0.01	–	–	0.01	0.22	0.71	0.07	$\leq 10^{-5}$

[a] From Cl2p core-level spectra. The Cl2p$_{3/2}$ BE for Cl$^-$ = 197.1 eV, Cl* = 198.6 eV and –Cl = 200.1 eV.
[b] From curve-fitted N1s core-level spectra.

Cl2p core-level spectra of the corresponding samples are shown in Fig. 10 [48]. The Cl2p core-level spectra of pristine and partially undoped complexes can be best resolved into three spin-orbit split doublets (Cl$2p_{3/2}$ and Cl$2p_{1/2}$), with the BE for the Cl$2p_{3/2}$ peaks lying at about 197.1, 198.6 and 200.1 eV. The first and the last BE components suggest the presence of ionic and covalent chlorine species, respectively. The intermediate chloride species, Cl* (dashed component in Fig. 10), on the other hand, can be more appropriately associated with anionic chloride species resulting from the CT interactions between the halogen and metal-like conducting state of the polymer chain. This intermediate chloride species was also observed in the Cl2p core-level spectrum of the HCl protonated

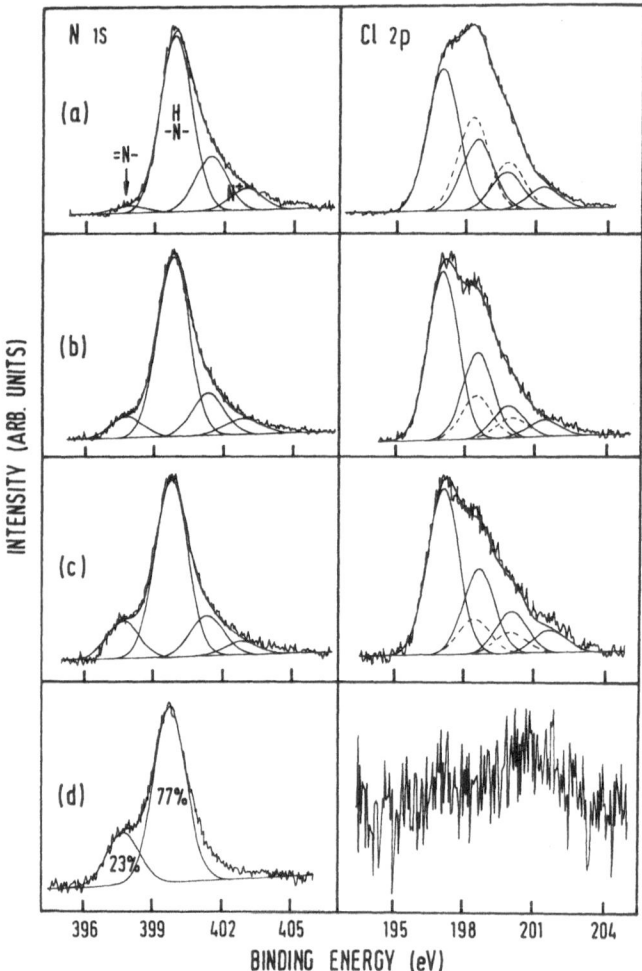

Fig. 10a–d. N1s and Cl2p core-level spectra of a PPY/chloride sample prepared in ethanol upon progressive NaOH treatments: (**a**) before treatment, Cl/N = 0.41; (**b**) Cl/N = 0.24; (**c**) Cl/N = 0.15; (**d**) Cl/N = 0.01

EM oxidation state of PAN [83, 84]. With progressive NaOH treatment, the Cl* species was preferentially removed over the Cl⁻ species by NaOH (compare the Cl*/N and Cl⁻/N ratios of Samples 1 and 2 in Table 3). The data in Table 3 further indicate that the decrease in the Cl⁻/N ratio, but not in the Cl*/N ratio, is accompanied by a corresponding increase in the proportion of the =N– component. These observations readily imply that each Cl⁻ species is associated with a specific nitrogen cation and that the Cl* species is less tightly bound. Nevertheless, a fairly close balance between the amount of chloride anions (Cl⁻ and Cl* species) and the proportion of pyrrolylium nitrogen cations is observed. Furthermore, the proportion of the neutral pyrrolylium nitrogens (–NH– species) remains fairly constant at all stages of the treatment. The residual N⁺ component in the completely undoped samples probably results from the replacement of anions by oxygen-containing species [85]. Thus, the presence of unit positively charged nitrogens implies that the polarons and bipolarons in PPY are more appropriately associated with the nitrogens so as to give a nitrogenonium ion polymer, as in the case of PAN [86] (see also below), but not with the carbon atoms, as is believed in general [87].

The presence of various chloride species in the PPY/chloride complex further suggests that any reaction stoichiometry based on the total chlorine balance of the complex can be quite misleading. For example, at least two reaction schemes which give rise to complexes which pyrrole:Cl⁻ ratios of 3:1 [88] and 4:1 [89] have been proposed. However, from the actual chloride anion content of the complex and the corresponding amount of deprotonated pyrrolylium nitrogens (~25%) after complete base compensation of the complex, it appears that the reaction stoichiometry corresponding to 25% oxidized pyrrole units in the complex is more appropriate.

Finally, it may be appropriate to refer to a recent XPS study on the PPY/ClO₄ complex treated with Cu(II) salt-containing alkaline solution [90]. This study found an adduct which gives a substantially broadened N1s core-level spectrum that can be resolved into two peak components with BE at about 399.1 eV and 397.5 eV. The lower BE component was attributed to imine nitrogens with covalently bonded Cu (N–Cu linkage) and also to their neighboring Cu-free pyrrolylium nitrogens. It is rather doubtful that efficient electron delocalization occurs between pyrrole units in such a way that the Cu-bonded nitrogen and its neighboring pyrrolylium nitrogen have an identical electronic environment, since PPY is basically a p-type conductor [21]. At this point, it may be relevant to note that for copper chloride complexed poly(2-vinylpyridine), the Cu-bonded pyridinium nitrogens undergo a positive shift in the N1s BE [91].

In the case of PAN, the nitrogenonium ion structure involving polaron and bipolaron defects has been associated with the protonated and highly conductive polymer [86]. Again, a number of earlier studies has suggested partially charged nitrogens or delocalization of positive charges [28, 29, 92]. On the other hand, the presence of unit positive charge on the nitrogen was favored by Munro et al. [78] on the basis of a similarity between the N1s BEs of the fully protonated PAN and the tetrabutylammonium perchlorate. The latter serves as a model compound having a unit positive charge on the nitrogen atom. Many more recent studies on the PAN-organic acceptor complexes [93] and PAN-halogen complexes [32, 94] have also concluded the presence of unit positively charged nitrogens from a critical comparison between the anion doping level and the proportion of positively charged nitrogens. For example, Table 4 shows the XPS results and stoichiometries obtained when fully reduced LM was increasingly oxidized by molecular bromine [32]. Since bromine is covalently bonded at high acceptor loading, there must be an optimum acceptor level for the LM/bromine complexes. However, a close balance between the number of bromide anions and that of positively charged nitrogens is observed at all acceptor levels. Table 5 shows the XPS results and stoichiometries of the LM/ o-bromanil complexes at various acceptor levels [94]. The oxidative doping of LM by an organic electron acceptor involves first the oxidation of the amine nitrogens and also the formation of the imine structure through hydrogen transfer from the pristine amine nitrogens to the acceptor. The imine nitrogens produced are then preferentially doped by the acceptor. These processes are revealed by the simultaneous appearance of the imine and the positively charged nitrogens, followed by the disappearance of the imine structure and a substantial increase in the amounts of positively charged nitrogens and halogen anions (and thus the electrical conductivity). Again, a close balance between the proportion of nitrogen cations and the anion/N ratio can be seen in each case.

Another important issue concerning the protonation of PAN appears to be whether the amine nitrogens of the EM oxidation state are also susceptible to protonation, although it has been well-established that protonation occurs preferentially at the imine units in EM [27]. An earlier report [22] predicted only a small ($\sim$ 3.6%) proportion of protonated amine sites at pH $\sim$ 0. On the otherhand, many other earlier XPS studies suggested protonation of both imine and amine nitrogens in the electrochemically prepared PAN, from the observed overall positive shift of the N$1s$ core-level BE [28, 76]. Theoretical calculations showed polymer structures with protonated amine units in the heavily oxidized

Table 4. XPS results and stoichiometries of various LM/bromide complexes

Sample	Bulk Br/N	Surface stoichiometry[a]			Proportion of			Conductivity (S/cm)
		Br/N	–Br/N	Br⁻/N	=N–[b]	–NH	N⁺	
1	0.05	0.11	0.01	0.10	0.08	0.74	0.18	5×10^{-6}
2	0.10	0.15	0.02	0.13	0.07	0.70	0.23	2×10^{-3}
3	0.30	0.27	0.03	0.24	0.07	0.60	0.33	5×10^{-1}
4	0.60	0.55	0.19	0.36	0.06	0.56	0.38	7×10^{-2}
5	1.0	0.65	0.31	0.34	0.03	0.62	0.35	1×10^{-2}
6	2.0	1.0	0.74	0.26	0.04	0.63	0.33	8×10^{-3}
7	4.4	1.31	1.10	0.21	0.03	0.69	0.28	2×10^{-3}

[a] Based on the corrected bromine to nitrogen core-level spectral area ratios.
[b] Neutral imine structure.

Table 5. XPS results and stoichiometries of LM/o-bromanil complexes

Sample No.	Bulk[a] o-bromanil/N ratio	Surface[b] o-bromanil/N ratio	Ratio[c] of		Proportion of			Conductivity (S/cm)
			Br⁻/N	-O⁻/N	-N=[d]	-NH-	N⁺	
1	0.04	0.04	0.02	0.04	0.07	0.83	0.10	3×10^{-5}
2	0.10	0.11	0.04	0.06	0.11	0.74	0.15	4×10^{-5}
3	0.20	0.24	0.07	0.11	0.11	0.72	0.17	4×10^{-5}
4	0.30	0.25	0.08	0.12	0.14	0.63	0.23	3×10^{-5}
5	0.50	0.40	0.17	0.13	0.04	0.64	0.32	2×10^{-2}

a Determined gravimetrically.
b Based on the corrected total halogen to nitrogen area ratio.
c Based on the anion peak component to total nitrogen area ratio.
d Neutral imine structure.

or doped PAN [95]. A more systematic study on the protonation of EM powder and film by H_2SO_4 and $HClO_4$ made it clear that a large proportion of the amine nitrogens is also susceptible to protonation, but only in the presence of excessive (acid/monomer ratio > 0.5) non-volatile acids [96]. Figure 11(a) to Fig. 11(d) compare, respectively, the N1s and Cl2p core-level spectra of two $HClO_4$-protonated EM films (cast from N-methylpyrrolidinone or NMP), with perchlorate/N ratios of about 0.5 and 0.78. Thus, for the highly protonated sample, the perchlorate/N ratio is consistent with the fact that about 74% of the nitrogens in the complex is positively charged and the proportion of the amine nitrogens decreases from about 50% to only 26%. Deprotonation of this complex by $0.5\ mol\,l^{-1}$ NaOH allows the polymer film to return to the EM oxidation state with about equal amounts of imine and amine nitrogens, not unlike that shown in Fig. 4(b). Therefore, at high $HClO_4$ loading, a substantial amount of the amine nitrogens must be protonated. Protonation of the poly-semiquinone form of PAN would give rise to $-N^+H_2-$ groups. This structure probably reduces the extent of π-conjugation or interrupts the polaron-bipolaron lattice [95], as is evident from the abrupt changes in the lineshape of the N1s high BE tail. Thus, the electrical conductivity of heavily protonated EM may actually fall below that of its 50% protonated counterpart.

Since the nitrogens in PPY and PAN are similar in chemical nature, it is of interest to compare the CT behavior of their corresponding intrinsic oxidation states. The DP-PPY and EM oxidation states exhibit very similar CT behavior

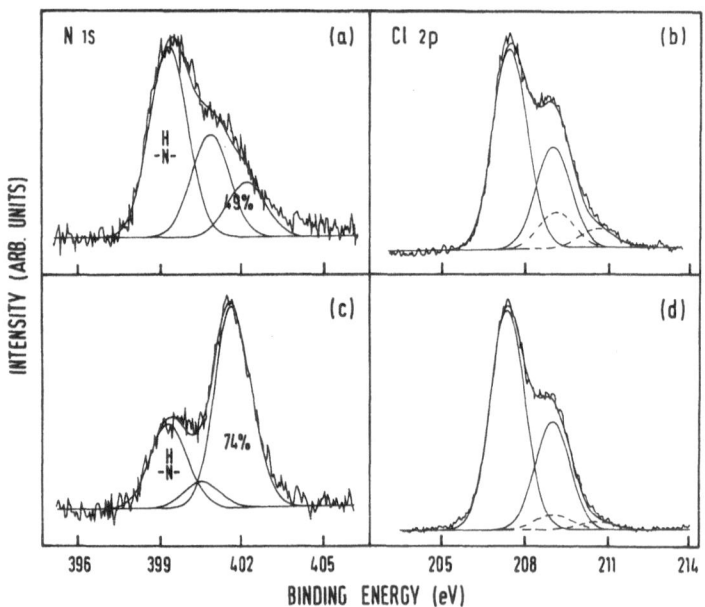

Fig. 11a–d. N1s and Cl2p core-level spectra of $HClO_4$ protonated EM films, with a perchlorate/N ratio of about 0.5 for (**a**) and (**b**), and 0.78 for (**c**) and (**d**)

towards organic electron acceptors, such as the halobenzoquinones [47]. In each polymer complex, the acceptor interacts preferentially with the imine units, as suggested by the reduction or disappearance of the imine component and the formation of the corresponding amount of the positively charged nitrogens in the N$1s$ core-level spectrum. The CT process is also accompanied by the formation of halogen and benzoquinone anions, as seen from their core-level spectra. The presence of the halogen anions in each complex implies that the CT interactions must have proceeded further than the formation of a pure molecular complex. Figure 12(a) to Fig. 12(d) show the N$1s$ and Cl$2p$ core-level spectra for the DP-PPY/o-chloranil and EM/o-chloranil complexes at acceptor/monomer ratios of 0.26 and 0.45, respectively. Table 6 compares the XPS data with the complex structures of several halobenzoquinone complexes. The plausible structures of DP-PPY and EM complexes, based on XPS data, are shown in Fig. 13, using the o-chloranil complex as an example in each case. The similarity of o-chloranil and o-bromanil in CT behavior towards the two base polymers is consistent with their similarity in chemical structure and reduction potential [97]. The substantially weaker CT interaction of the two polymers with p-chloranil supports a complex structure involving the cleavage of the carbon-halogen bond and the subsequent formation of a new linkage through the C_4 position of the acceptor. This position is sterically less hindered and exists only in $ortho$-halobenzoquinones. In general, a substantially higher extent of CT is observed between the two base polymers and the $ortho$-halobenzoquinones [82, 98].

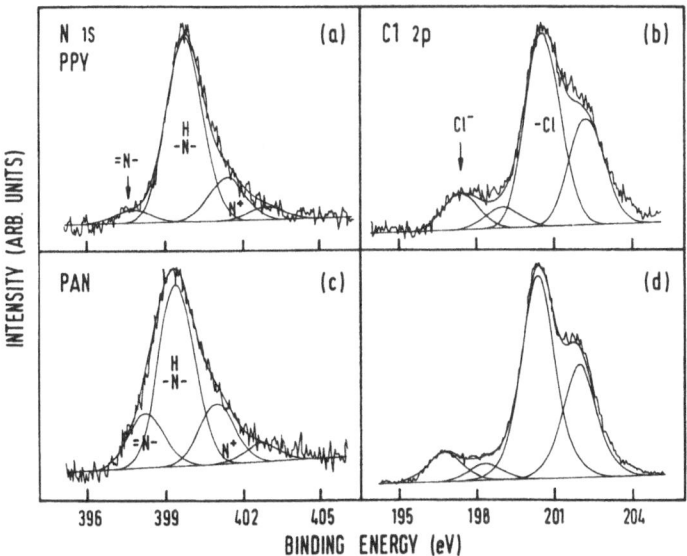

Fig. 12a–d. N$1s$ and Cl$2p$ core-level spectra for (**a**) and (**b**) DP-PPY/o-chloranil complex (acceptor/monomer = 0.26); (**c**) and (**d**) EM/o-chloranil complex (acceptor/monomer = 0.45)

Table 6. XPS results and stoichiometries of various organic acceptor complexes of DP-PPY and EM base

Complex	Acceptor/Monomer ratio[a]	Proportion of			Anion/N ratio		Conductivity (S/cm)
		$-N=$	$-NH-$	N^+	Halogen/N	O/N	
DP-PPY/o-chloranil	0.26	0.05	0.73	0.22	0.16	0.10	3×10^{-1}
EM/o-chloranil	0.45	0.17	0.57	0.26	0.20	0.10	3×10^{-2}
DP-PPY/o-bromanil	0.18	0.12	0.72	0.16	0.09	0.10	2×10^{-2}
EM/o-bromanil	0.30	0.27	0.50	0.23	0.16	0.12	4×10^{-3}
DP-PPY/p-chloranil	0.16	0.13	0.71	0.16	0.03	0.12	8×10^{-3}
EM/p-chloranil	0.29	0.34	0.47	0.19	0.13	0.06	6×10^{-5}

[a] Determined from the corrected nitrogen and halogen core-level spectral area ratios and agreed to within $\pm 10\%$ of the bulk composition.

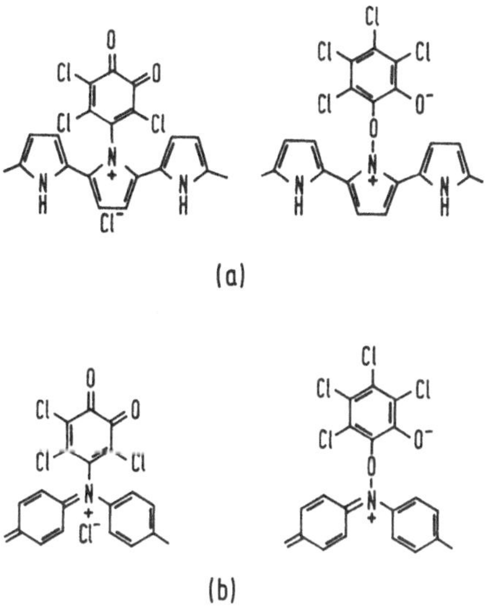

(a)

(b)

Fig. 13a, b. Plausible structures of (a) DP-PPY/o-chloranil and (b) EM/o-chloranil complexes

Another interesting area of CT studies involves the self-doped or self-protonated PAN and PPY. Yue and Epstein [99] first reported on the synthesis and XPS study of sulfonic acid ring substituted and self-protonated PAN. With the chemical similarity of the nitrogens between PAN and PPY taken into consideration, sulfonated and self-doped conducting PPY was also prepared [100]. Figure 14(a) to Fig. 14(f) show the N$1s$ core-level spectra of EM, LM and DP-PPY bases before and after treatment with fuming H_2SO_4. The S/N ratios for the sulfonated EM, LM and DP-PPY complexes are 0.47, 0.48 and 0.23, respectively. This ratio for the DP-PPY complex is only one half of that of its EM counterpart in PAN, consistent with the fact that the intrinsic oxidation state or the proportion of the =N− units in DP-PPY is only one half of that of the the EM base. The disappearance of the imine component in the N$1s$ core-level spectra of sulfonated EM and DP-PPY implies that self-protonation also occurs preferentially through the imine units. The self-protonation or self-doping is evident from the appearance of the high BE tail above 400 eV, characteristic of positively charged nitrogens, in the N$1s$ core-level spectra. For each complex, the S/N ratio agrees fairly well with the proportion of the positively charged nitrogens.

For sulfonated DP-PPY, the appearance of a new low BE component, which is shifted by only about − 1.1 eV from the neutral pyrrolylium nitrogens and is to be distinguished from the imine nitrogen component, is associated with the nitrogens in the sulfonated ring. Thus, the simultaneous appearance of this low BE N$1s$ component and the protonated imine nitrogens during sulfonation implies that sulfonation and protonation in DP-PPY occur on different pyrrolylium rings. Figure 14(g) to Fig. 14(i) show the N$1s$ core-level spectra for the sulfonated EM, LM and DP-PPY complexes after 'undoping' with the stoichiometric amount of NaOH. The S/N ratio for each complex remains unchanged

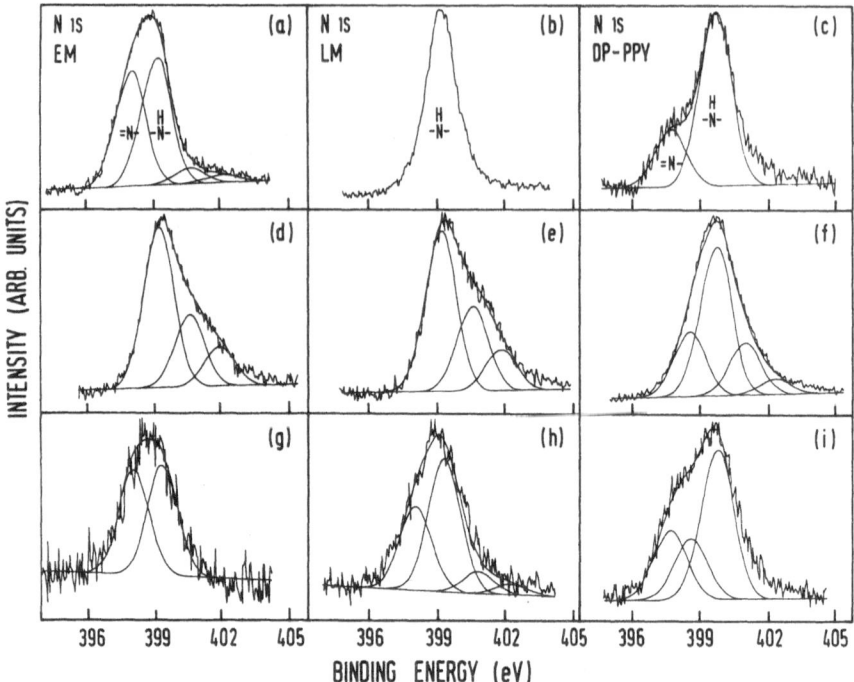

Fig. 14a–i. Nls core-level spectra of (**a**) EM; (**b**) LM; (**c**) DP-PPY; sulfonated and self-doped (**d**) EM (S/N = 0.47), (**e**) LM (S/N = 0.48), (**f**) DP-PPY (S/N = 0.23); sulfonated Na salts of (**g**) EM, (**h**) LM, (**i**) DP-PPY

during the undoping process. The Nls core-level spectra of the EM and DP-PPY complexes reveal that base treatment returns the polymers to their respective intrinsic oxidation states. In this connection, it may be appropriate to note that Yue and Epstein [99] reported an Nls spectrum of the Na salt of sulfonated EM base which has a linewidth and an amine/imine peak separation substantially smaller than those of the pristine EM base. Thus, substantial structural changes might have occurred in their sample during the base treatment. The persistence of the peak component at about 398.6 eV in the Na salt of

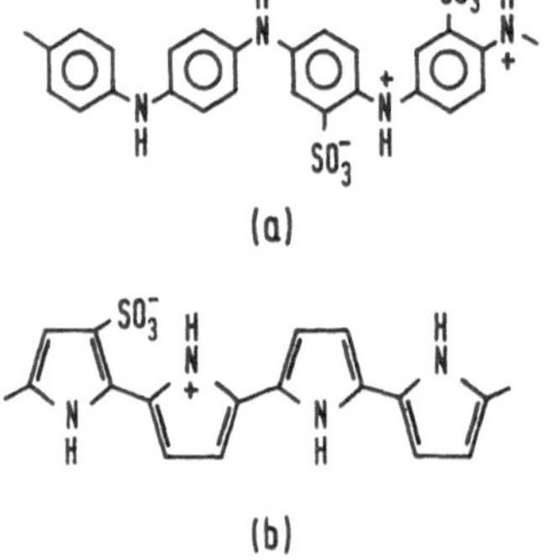

Fig. 15a, b. Plausible structures of sulfonated and self-protonated (**a**) EM and (**b**) DP-PPY

sulfonated DP-PPY confirms the earlier conclusion that sulfonation and protonation in DP-PPY occur on separate pyrrolylium units. Base treatment of sulfonated LM gives rise to a substantial proportion of imine nitrogens. The structure of sulfonated and self-doped LM, therefore, should be essentially similar to that of its sulfonated EM counterpart. Figure 15 shows the plausible structures of sulfonated and self-protonated EM and DP-PPY derived from the XPS results.

3.2.1.3 S-Containing Polymers

The thiophene and related families of polymers can be synthesized either chemically or electrochemically. However, most of the XPS studies have focused on complexes synthesized by the latter method. The earlier work of Hotta et al. [101] demonstrated that XPS provides a convenient tool for the determination of both the doping level and dopant species in poly(3-methylthienylene). Considerable discrepancies can be found in the subsequent XPS studies on PTH and its alkyl-substituted complexes. The most important issues are probably the charge distribution and the structure of the oxidized or doped PTH chain.

Many studies [102, 103] have suggested charge withdrawal to occur only from the ring carbon atoms, on the basis of the observed positive BE shift in the $C1s$ core-level spectrum and essentially no chemical shift in the $S2p$ core-level spectrum for doped PTH. The study of Takenaka et al. [104] on the PTH/BF_4 complexes focused mainly on the shift in the $C1s$ BE. On the other hand, Riga et al. [55] argued that in the oxidized state, both the carbon and sulfur atoms are positively polarized, as indicated by the fact that both the $C1s$ and $S2p$ core-level spectra show an overall $+ 0.7$ eV shift in BE. It was further suggested that because the $S2p$ chemical shift is less sensitive to the charge than the $C1s$ shift, the identical values observed in the XPS shifts actually correspond to a larger charge extraction from the sulfur atoms. Yet another study [105] found BE shifts of the order of about 0.4 eV in the $S2p$ core-level spectra for doped poly(alkylthiophene)s exposed to air or subjected to thermal undoping.

The lack of consistency in the reported XPS data on thiophene polymers is at least partially attributable to little or no attempt having been made to resolve the $S2p$ core-level spectrum. $S2p$ spectral broadening has been reported for the perchlorate doped PTH, PBiT and PMeT synthesized electrochemically [56]. Figure 16(a) and Fig. 16(b) show the $C1s$ and $S2p$ core-level spectra of a PBiT complex synthesized chemically in the presence of $Cu(ClO_4)_2 \cdot 6H_2O$ in acetonitrile [106]. The complex has a perchlorate/S ratio of 0.25, as determined from the corrected $Cl2p/S2p$ core-level spectral area ratio. The $S2p$ core-level spectrum is best resolved into two major spin-orbit split doublets ($S2p_{3/2}$ and $S2p_{1/2}$), with the BE for the $S2p_{3/2}$ peaks lying at about 163.6 eV and 164.6 eV. The former is attributable to the neutral thiophene units [55]. The smaller but distinct high BE component (dashed curves) which is shifted by about $+ 1.0$ eV from the neutral sulfur species has been associated with the positively polarized or partially charged sulfur species that appears after charge extraction from some thiophene units by the dopant. No BE shift of the main C1s peak component was observed for the oxidized PBiT complexes, even after an extended exposure to air. Instead, an asymmetric C1s spectrum skewed towards the high BE side was observed for the doped complexes. The high BE C1s components, the proportion of which is comparable to the proportion of the high BE S2p component, again can be attributed to the positively polarized carbon atoms of the positively charged thiophene units. The high BE tails in the C1s (BE > 288 eV) and S2p (BE > 167 eV) core-level spectra of the complex are due to the shake-up satellite structure ($\pi \rightarrow \pi^*$ transition, Fig. 16).

The assignment of the high BE components in the C1s and S2p core-level spectra to the positively polarized species is further supported by the gradual

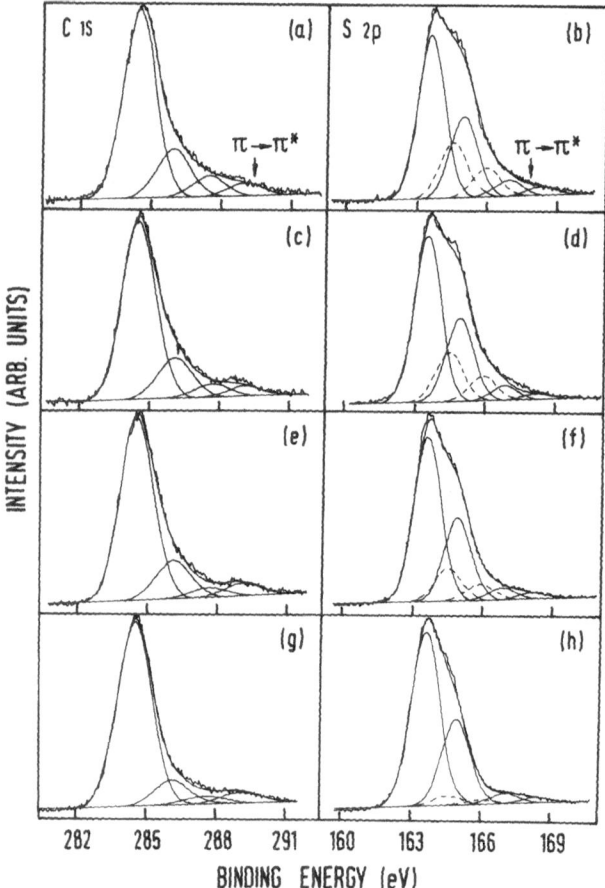

Fig. 16a–h. C1s and S2p core-level spectra of a PBiT/perchlorate complex synthesized in the presence of $Cu(ClO_4)_2 \cdot 6H_2O$ at various extents of undoping by NaOH, (**a**) and (**b**) pristine sample, perchlorate/S = 0.25; (**c**) and (**d**) perchlorate/S = 0.18; (**e**) and (**f**) perchlorate/S = 0.07; (**g**) and (**h**) perchlorate/S ≈ 0.0

Table 7. Effect of varying extent of NaOH treatment on a PBiT/prechlorate complex synthesized from $Cu(ClO_4)_2 \cdot 6H_2O$ in acetonitrile

Sample	Complex stoichiometry[a] perchlorate/S	Chemical states		Proportion[b]		Proportion[b]		Conductivity (S/cm)
		ClO_4^-/S	ClO_4^*/S	S^+	S^0	C^+	C^0	
1 (Pristine)	0.25	0.20	0.05	0.24	0.76	0.25	0.75	1
2	0.18	0.15	0.03	0.22	0.78	0.23	0.77	10^{-1}
3	0.07	0.07	0.0	0.15	0.85	0.18	0.82	10^{-4}
4	0.0	0.0	0.0	0.04	0.96	0.14	0.86	$< 10^{-10}$

[a] Based on the corrected Cl2p and S2p core-level spectral area ratios.
[b] S^+ and C^+ are polarized or partially charged species while S^0 and C^0 are neutral species.

disappearance of these components with progressive undoping of the polymer complex by NaOH, as illustrated in Fig. 16(c) to Fig. 16(h). The complex stoichiometries and the proportions of various carbon and sulfur species at various extents of undoping are summarized in Table 7. Thus, for the more fully oxidized complexes, which are less susceptible to contamination due to surface oxidation, a fairly close balance is always observed between the dopant anion/monomer ratios and the proportions of positively polarized carbon or sulfur atoms. Most important of all for the completely undoped sample is that the C1s core-level spectrum becomes fairly symmetrical and the high BE component in the S2p core-level spectrum at 164.6 eV disappears almost completely. Similar undoping behavior was also observed in alkyl substituted PTH, such as PMeT, complexes. Thus, in the oxidized thiophene polymer complexes, the dopant anion is associated with a specific thiophene ring and each positive charge is localized on the corresponding thiophene ring to form the thiophenium ion. This conclusion is consistent with the simultaneous presence of neutral and appropriate proportions of positively polarized carbon and sulfur species. The formation of the thiophenium ion, as suggested by the presence of similarly well-resolved C1s and S2p high BE components, has also been reported in a recent XPS study on electrochemically oxidized PBiT [107].

A more extensive charge localization on the sulfur heteroatoms, as suggested by a large BE shift for positively charge sulfur species, was observed for doped PPS. The S2p core-level spectrum of TaF$_5$-doped PPS showed a new high BE component, which shifted by + 2.7 eV from the neutral sulfur species of PPS at 163.3 eV [108, 109]. An S2p BE shift of comparable magnitude was also found in PPS heavily doped with SbF$_5$ [110]. This BE shift is much larger than the + 1.0 eV shift observed in the thiophenium ion. The simultaneous broadening of the C1s spectrum suggests that the positive charges are delocalized over the corresponding phenyl rings. These facts are consistent with the formation of a more planar PPS chain that occurs as a result of an increase in the double bond character of C–S bond upon doping [109]. In the presence of a weaker acceptor, the C1s and S2p spectral broadenings much diminish owing to reduced CT. For example, for SO$_3$-doped PPS, the chemical shifts for the high BE C1s and S2p components reduce to only about 0.5 eV. No spectral broadening was observed for PPS doped with iodine or TCNQ [110].

3.2.1.4 Other Conjugated Polymers

For PPV doped with AsF$_6^-$ anion, a new high BE C1s component appeared at about 286.5 eV, in addition to the two C1s components at 284.6 eV and 285.5 eV for pristine PPV [60]. It has been associated with carbon atoms in a highly oxidized state resulting from doping. Thus, the C1s core-level spectrum of doped PPV readily indicates that the positive charge is non-uniformly distributed over the polymeric matrix. Furthermore, the satellite structures of the doped polymer are found to have a substantially enhanced intensity at about 290 eV, which is nearly 2 eV lower than that of the neutral polymer. This increase in satellite intensity has been interpreted as due to the deformation of valence molecular orbitals of the polymer matrix in the polar medium.

3.2.2 Non-Conjugated Polymers

Vinylpyridine polymers are among the most widely studied electroactive polymers containing non-conjugated backbones. The poly(2-vinylpyridine)/iodine ($P2VP/I_2$) complexes have found applications in solid-state electrochemical cells [111]. Semiconductive polymer complexes have also been made by using P2VP and poly(4-vinylpyridine) (P4VP) as donor polymers and iodine or halobenzoquinones as acceptors [112, 113]. XPS core-level spectra are again useful for the elucidation of the CT processes and the resulting complex structures in vinylpyridine polymers. From the proportion of positively charged pyridinium nitrogens and the amount of anions present in the CT complex, the steric hindrance associated with the pyridinium nitrogen at the *ortho*-position of P2VP has been found to have a marked effect on the extent of CT with bulky acceptors, such as DDQ. Furthermore, it was found that *ortho*-halobenzoquinones, such as *o*-chloranil and *o*-bromanil, are more efficient acceptors than their *para*-counterparts [114]. This finding can be interpreted as due to the involvement of the sterically less hindered C_4 position of the acceptor during the CT process. Similar effects were also observed in the organic complexes of PPY and PAN discussed in Sect. 3.2.1. The formation of the halogen and benzoquinone anions and the conversion of an appropriate portion of the neutral pyridinium nitrogenes to positively charged species suggest that the structures of the CT complexes involving vinylpyridine polymers and organic acceptors are approximately similar to those shown in Fig. 13.

XPS has also been proven useful for elucidating the ligand structure formed between the nitrogens of P2VP and copper chloride [91]. The appearance of a new high BE component positively shifted by about 1 eV from the neutral pyridinium nitrogens in the $N1s$ core-level spectrum of the complex has been attributed to a partial CT from the nitrogen lone pair to copper. The complexation of only one pyridine ligand by Cu(II) has been considered as due to the steric hindrance encountered in P2VP. At least two pyridine moieties of sterically less hindered P4VP can complex with each Cu atom to give a cross-linked polymer. According to these XPS data, the complex is thermally decomposed to form metallic copper, first by a complete reduction of Cu(II) to Cu(I). This is suggested by the change in $Cu2p_{3/2}$ BE from about 935 eV to about 932.8 eV. The Cu(I) state can be distinguished from the metallic Cu not from their $2p_{3/2}$ BEs, but their Auger parameters (the difference in KE between an Auger line and a photoelectron line) are 1847.8 ± 0.2 eV for the former and 1851.1 ± 0.3 eV for the latter. Thus, the P2VP/copper chloride complex represents a soluble and processable polymer/metal complex whose thermal degradation behavior offers potential applications in the area of direct formation of conductive paths for electronic circuits.

Vinylcarbazole polymers and their CT complexes are widely known for their excellent photoconductive properties [115]. The CT processes in this polymer system have been extensively worked out by the conventional absorption spectroscopic techniques, but the application of the XPS technique to them has been rather limited [116]. For poly(N-vinylcarbazole) (PVK)/perchlorate complex film synthesized via electrochemical polymerization and oxidation, it was shown that the N$1s$ core-level spectrum exhibits a new high BE component shifted by about $+3.0$ eV from the neutral carbazole nitrogen at about 400 eV. This new component was attributed to the positively charged carbazole nitrogen associated with ClO_4^- anion [117].

3.3 Stability and Degradation

In considering the potential applications of electroactive polymers, the question always arises as to their stability. The deterioration of a physical property such as conductivity can be easily measured, but the chemical processes underlying it are not as easy to be revealed. In order to understand them, XPS has been used to follow the structural changes which occur in the polymer chain and the counter-ions of the doped polymer. The following sections present some XPS findings on the degradation of electroactive polymers, such as polyacetylene, polypyrrole, polythiophene and polyaniline, in the undoped and doped states.

3.3.1 Undoped Polymers

The prototype of electrically conducting polymers is $(CH)_x$, which is an insulating material as polymerized and becomes conductive upon doping. The as-polymerized material reacts with oxygen in air, yielding in the conductivity first increasing to a maximum and then decreasing [117, 118]. The reaction with oxygen has been proposed [119–121] to be a 2-stage phenomenon: the first stage involves the reversible formation of a $(CH)_x$/oxygen CT complex, which promotes an increase in conductivity, while the second stage is an irreversible reaction involving the destruction of the polyene carbon–carbon double bond which leads to a loss of conductivity. The XPS investigations by Munstedt [17] clearly show an increase in the intensity of the high BE tail in the C$1s$ core-level spectrum of the $(CH)_x$ after exposure to air for extended periods (Fig. 17). The deconvolution of the high BE tail gives rise to three peaks which can be assigned to the oxygenated species C–O, C=O and O–C=O. The data indicate that the proportions of carbonyl and ester groups increase faster than that of the ether group, although the latter constitutes the major fraction of the oxygenated species at the initial stages (Fig. 3). The presence of such species have also been confirmed by infra-red (IR) absorption spectroscopy [118].

Due to its high sensitivity to oxygen, practical applications of $(CH)_x$ may be limited. Polyacetylenes with various substituents have been synthesized, and

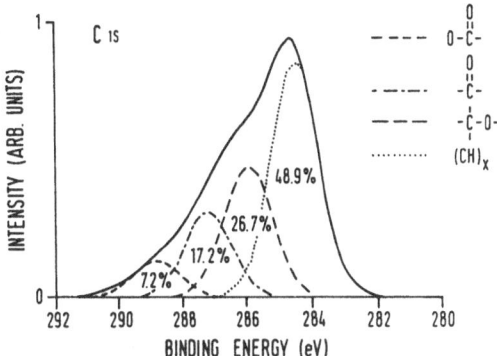

Fig. 17. $C1s$ core-level spectrum of $(CH)_x$ film after exposure to air for 22 days

their stability in air and thermal degradation behavior have been studied [20, 122]. In general, polyacetylenes with substituents are thermally more stable than $(CH)_x$ in air, with their stability increasing with increasing number and/or bulkiness of the substituents [20]. Of particular interest are PPA and Poly-$(o\text{-}Me_3SiPA)$, both of which possess good photoconductive properties [74, 75]. At room temperature, PPA also undergoes reaction with oxygen to some extent as shown by the presence of oxygen containing groups in the $C1s$ core-level spectra [123]. Upon heating to 130 °C in O_2, a weight gain takes place and the intensity of the high BE tail in the $C1s$ spectrum increases appreciably. Deconvolution of the $C1s$ spectrum indicates the proportion of C–O structure is about seven times as much as that of the C=O groups. Elemental analysis shows similarity of the heated sample to the pristine one in the H/C ratio, thus suggesting that the formation of oxygen containing groups is primarily the addition to the double bonds in the chain [123]. In contrast, the $C1s$ and $Si2p$ core-level spectra of Poly($o\text{-}Me_3SiPA$) after heating to 270 °C in a 50% O_2–50% N_2 mixture are similar to those of the original sample. The absence of oxygenated species such as C=O, C–O and Si–O groups is consistent with the absence of absorption bands due to these groups in the IR absorption spectra [124].

Another factor which may also limit the potential applications of $(CH)_x$ is its insolubility in organic solvents. On the other hand, both PPA and Poly($o\text{-}Me_3SiPA$), as well as many other substituted acetylene polymers [20], can be cast into film from common organic solvents. However, it has been found that PPA undergoes a gradual loss of effective conjugation and a decrease in molecular weight when dissolved in organic solvents [125]. The degradation rate is affected by choice of solvent, temperature, solution concentration, light illumination as well as the microstructure of the polymer. Poly($o\text{-}Me_3SiPA$) has been found to be highly stable in organic solvents in the absence of light. When irradiated with ultra-violet (UV) and short-wavelength visible light in chlorinated solvents, it degraded at a rate faster in more chlorinated solvents. The $Cl2p$ core-level spectrum of the degraded sample recovered from a Poly$(o\text{-}Me_3SiPA)/CCl_4$ solution [126] showed a main peak centered at about

200.5 eV, which corresponds to the $Cl2p_{3/2}$ BE of chlorine bonded to C atoms, along with a much smaller peak at 199.2 eV. The latter peak is an indication of the formation of a CT complex between chlorine and the polymer. When $CHCl_3$ was used, the $Cl2p$ core-level spectrum of the degraded sample showed a lower BE component which is about 66% of the covalent chlorine component.

The stability problem with $(CH)_x$ has directed a greater interest towards other conjugated polymers such as PPY and PTH. These polymers can be synthesized by either electrochemical or chemical polymerization and oxidation of the respective monomers. The as-prepared polymer complexes can be un-doped electrochemically or chemically with the removal of the counter-ions. As discussed earlier, XPS studies on chemically synthesized PPY have shown that when deprotonated with NaOH, the polymer base (DP-PPY) contains about 20–25% oxidized imine-like (–N=) structure which can be reduced to the all amine-like (–NH–) structure of PPY°. From the $N1s$ core-level spectra of DP-PPY and PPY°, it was found that a persistent high BE tail is present [127], which can partially be attributed to surface oxidation products, since un-complexed PPY has a relatively low oxidation potential [85, 128, 129]. Pfluger et al. [130] showed that undoped PPY irreversibly absorbed large amounts of oxygen with a 10^4 increase in conductivity after absorption of 1% by weight of O_2. According to their XPS data, chemical reactions occur at the nitrogen atoms but can be reversed electrochemically [130].

In the case of PAN, when light grey, fully reduced LM powder is exposed to air, it gradually darkens. The possibility that some of its amine units undergo oxidation to imine units in air has been suspected [22]. The $N1s$ core-level spectra of a freshly prepared LM sample and the sample after exposure to dry air for two months are compared in Fig. 18(a) and 18(b) [131]. As expected, the former exhibits only a single nitrogen environment at a BE of 399.3 eV which is characteristic of the amine structure. After exposure to air, the spectrum is skewed slightly towards the low BE side, and a new peak component at 398.1 eV, characteristic of the imine units, is resolvable. This finding confirms the oxidation of some amine units to imine units, but the process appears to go rather slowly under ambient conditions. Thermogravimetric scans of LM heated from 25–700 °C in N_2, air and a 50% O_2–50% N_2 mixture showed no significant weight changes below 300 °C [131]. However, XPS analyses of LM samples after heat treatment at 270 °C in both N_2 and 50% O_2–50% N_2 indicated that these samples undergo transformation, resulting in the formation of imine-type species (Figs. 18(c) and 18(d)). As expected, the fraction of nitrogen existing as such species is much higher in the latter sample. The increase in imine-type structure upon heat treatment as indicated by the $N1s$ core-level spectra is supported by the IR absorption spectra which show an increase in the intensity of the $1600\ cm^{-1}$ band attributable to the quinonoid ring relative to that of the $1500\ cm^{-1}$ band attributable to the benzenoid ring [132]. Further-more, elemental analysis of the heated samples showed a decrease in the H/C ratio. In Figs. 18(c) and 18(d), a high BE component at about 401 eV is resolvable. This is characteristic of positively charged nitrogen which may be

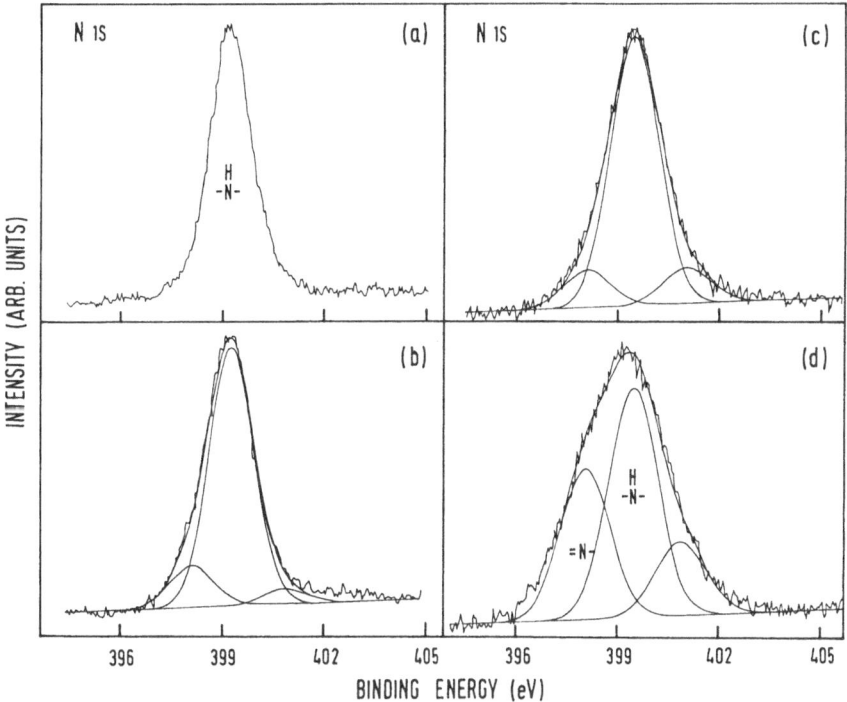

Fig. 18a–d. N$1s$ core-level spectra of (**a**) freshly prepared LM; (**b**) LM after two months' exposure to dry air at 25 °C; (**c**) LM after heating in N$_2$ at 270 °C for 2 h; (**d**) LM after heating in 50% O$_2$–50% N$_2$ at 270 °C for 2 h

ascribed to surface oxidation products. The presence of a significant amount of positively charged nitrogen even in a sample heated in N$_2$ suggests reaction with O$_2$ in air during sample handling after the heat treatment.

The solubility of PAN in NMP enables free-standing film to be cast from the solution [23]. However, UV-visible absorption spectroscopy of LM in NMP suggests that dissolved oxygen oxidizes the amine units. The oxidation is accelerated by UV-visible light irradiation [133]. The XPS analysis of the solid recovered from the NMP solution confirms the presence of imine units. In the case of the 50% intrinsically oxidized PAN (EM base) dissolved in NMP, the optical spectrum of the solution kept in air shows no significant changes for several days. However, after 1h of UV-visible light irradiation, the UV-visible absorption spectrum shows clear changes in the band intensity and position. A significant broadening of the C$1s$ core-level spectrum and a distinct shake-up structure are observed. These features indicate the presence of oxygen containing groups such as C=O or C–O and a more localized electronic structure responsible for a decrease in effective conjugation. However, no further oxidation of the amine units of EM base to imine units has been detected by XPS.

It has been claimed that fully oxidized PNA can be obtained by oxidation of the EM base by ammonium persulfate or by compensation of heavily acceptor-doped EM [35]. Evidence for such PAN was obtained from UV-visible and IR absorption spectroscopic data. However, XPS analysis of PAN samples obtained from the compensation of heavily iodine-doped LM or EM base samples showed that about 25% of the nitrogen still remain as amine units [32], although their IR and UV-visible absorptions were similar to those described in Ref. [35]. Thus, XPS appears to be a more reliable and quantitative technique for analyzing the oxidation states of PAN. Snauwaert et al. [33] showed that the maximum oxidation state of PAN achievable electrochemically using aqueous acid electrolytes was 75% (i.e. containing 75% imine nitrogens), when deter-mined from the $N1s$ core-level spectra. The use of a higher potential leads not to further oxidation but to the hydrolysis of the polymer.

Although PPY and PTH are not soluble in common organic solvents including NMP, soluble PPY [134, 135] and PTH derivatives [136, 137] were synthesized. In particular, alkyl-substituted PTHs have been subjected to extensive study because of their high processability [57, 137, 138], and poly(3-alkylthiophene)s were found to exhibit thermochromism [52, 57, 138–140]. Motivated by this finding, a series of studies on the temperature dependence of the electronic structure of undoped poly(3-hexylthiophene) or PHeT were carried out using XPS and UPS [52, 57, 133]. Thus, from the XPS $C1s$ core-level spectrum, it was concluded that the presence of shake-up peaks at higher temperature reflects a more localized electronic structure, while the disappearance of the peaks at low temperature evidences a stronger delocaliza-tion. The UPS spectra revealed a narrowing of the lowest BE π band, which is consistent with the enhancement of π-electronic localization with increasing temperature. The shake-up features of model molecules were compared to those of poly(3-alkylthiophene)s, and an analysis of the gas-phase molecular data confirmed the geometric model of electronic localization in the polymers [57]. Based on the similarity between the electronic structure in thin solid films at low temperature and that in solution in poor solvents, the solvatochromic effects have been rationalized with the concept of conformational defects.

Undoped PPP and PPS (oxidation potential, $E_{ox} \simeq 1.6$ V) have a higher oxidation stability than undoped PPY ($E_{ox} \simeq -0.3$ V) [141]. XPS was used to investigate the extent and nature of the oxidation of PPS powder under slurry conditions, either in toluene at 60 °C with a mixture of formic acid and hydrogen peroxide as the oxidizing agent or in methylene chloride at 25 °C with 3-chloroperoxybenzoic acid [142]. The results indicated that up to 75% of sulfide (–S–) sulfur is oxidized to sulfoxide (–(S=O)–) and sulfone (–(O=S=O)–) in the surface region. The oxidation did not appear to affect the phenyl ring, and both unoxidized and oxidized PPS surfaces showed the $\pi \rightarrow \pi^*$ transition associated with the unsaturation in the phenyl ring. Depth profiling with Ar^+ ion sputtering exhibited an increase in the sulfide concentration and a decrease in the sulfone concentration. These findings suggest that the core of the PPS polymer particle remains unchanged under the oxidizing conditions studied.

3.3.2 Doped Polymers

In the earlier sections, we referred to the high affinity of undoped $(CH)_x$ for oxygen and electron acceptors. The stability and degradation of a doped polymer will depend on its reactions with the counter-ion or the constituents of the environment. In addition, the reactions of the counter-ions with the latter must be considered. Chiang et al. [5, 143–145] have shown that the electrical properties of $(CH)_x$ films undergo a dramatic change when doped with electron acceptors such as halogens and AsF_5 or with electron donors such as sodium napthalide in a THF solution. In the case of AsF_5-doped $(CH)_x$, the conductivity decreased with time when the sample was heated above $50\,°C$ in vacuum [146]. This decrease was accompanied by an appreciable decrease in the integrated $F1s$ XPS peak intensity (Fig. 19), whereas the magnitude of the $As3d$ peak decreased less drastically. However, the As peaks shifted towards lower BE, and this was taken as due to reduction to metallic As. The $C1s$ peak showed no additional structure indicative of bonding between the carbon and arsenic fluoride. Based on the combined XPS and mass spectroscopy data, the following decomposition path for AsF_5-doped $(CH)_x$ has been proposed:

$$4[(CH)_x^+(AsF_6)^-] \longrightarrow 4(CH)_x + 2As + 2AsF_3 + 9F_2$$

The AsF_3 and F_2 molecules were desorbed as gases into the vacuum chamber and As was left in the matrix of $(CH)_x$ film. For electrochemical doping, it has been found that doping levels above 8 mole % led to an irreversible conductivity loss, with IR evidence of C–F bond formation. Between 1 to 7 mole % dopant, the conductivity was stable for several weeks in dry air since the polymer no longer reacts with oxygen [118].

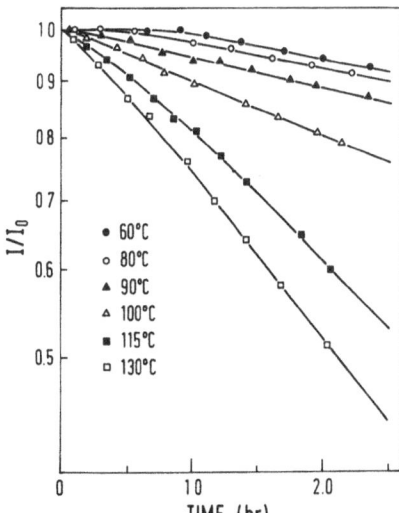

Fig. 19. Decrease in integrated $F1s$ peak intensity of AsF_5-doped $(CH)_x$ upon heating at different temperature in vacuum. I_0 is the magnitude of the initial integrated $F1s$ peak

Raman spectroscopy [147] and XPS [16, 63, 64] on iodine-doped $(CH)_x$ films identified I_5^- and I_3^-. In addition to these two active dopant species, a large amount of I_2 was found in films doped to saturation. XPS studies have also shown from the $I3d_{5/2}$ core-level spectra that the intensity of the I_5^- peak (at a BE of 620.6 eV) decreases relative to the I_3^- peak (at BE of 619.0 eV) during thermal treatment of the doped film. Similar observations have been made using Raman spectroscopy [148]. These observations and mass spectroscopy results led Österholm et al. [149] to the following thermal decomposition mechanism for $(CH)_x$ films highly doped with iodine. Initially, the dopant is desorbed at a fast rate owing to a simultaneous desorption of I_2 residing loosely on the surface of the fibrils and of I_2 produced by the decomposition of the thermally less stable I_5^- into I_3^- and I_2. After longer heating the rate of iodine decomposition is slowed down because the rate of formation of I_2 is reduced since most of the I_5^- species has already been decomposed. At this stage, the film is almost entirely doped by I_3^- species, and some of I_3^- will slowly decompose into I_2. Iodine-doped film is less stable than AsF_5-doped film, since the decomposition commences above room temperature in vacuum.

Other stability studies [150–152] have shown that iodine as well as perchlorate-doped samples lose conductivity quite rapidly in vacuum owing to the reaction of the polymer with the counter-ions. It has also been proposed that the decay in the initial high conductivity of iodine-doped $(CH)_x$ under argon atmosphere is due to secondary reactions involving the I_3^- and I_5^- counter-ions with the mid-gap states on the polymer backbone [153]. Another proposal was that in aqueous environments, reactive hydroxyl anions similarly reacting with these mid-gap states are partially responsible for the accelerated degradation [153]. Stability studies were carried out on $(CH)_x$ doped by a variety of other dopants such as BF_4^- [17], HCl [154], HI and HBr [155] and $FeCl_3$ [156], and the effect of O_2 and H_2O on conductivity was also investigated. However, in most cases, owing to the lack of structural data from XPS, reactions and degradation products have not been determined.

In the case of PPY, although the electrochemically oxidized state is more stable to atmospheric exposure than is the undoped state, a decrease in conductivity as well as an optical change occurred during a prolonged storage of the former in air [128, 157–159]. When PPY is doped with BF_4^-, it is likely that the anion originally incorporated into the film is also replaced by some oxygen-containing species during repeated electrochemical switchings [77, 160]. XPS has been used to study the reactions of electrochemically prepared BF_4^--doped PPY exposed to various atmospheres at room temperature [161]. Little conductivity change was observed on exposure to dry or water-saturated argon over 500 h. However, when exposed to dry air or oxygen or water-saturated oxygen, a measurable conductivity decay was found. The most significant effect occurred with water-saturated oxygen, leading to a decrease in conductivity by a factor of 7 over 500 h. In all cases, irrespective of the composition of the storage atmosphere, XPS results showed a decrease in the concentration of BF_4^- anions which cannot be directly correlated with the conductivity changes. It has been suggested that water incorporated in the polymer during or immediately after preparation is involved in a process leading to the anion depletion [161].

It has also been reported that electrochemically reduced PPY after exposure to air gives rise to an N$1s$ peak that has a low BE shoulder [160, 162], similar to that observed in NaOH-treated samples. This effect can be reversed upon treatment with HCl [42, 162]. The treatment of BF$_4^-$-doped PPY with oxygen-saturated water also yielded a similar structure in the N$1s$ spectrum [161], but with oxygen-free water, the intensity of the low BE shoulder was much weaker. In both cases, the conductivity decreased, but its value for oxygen-free water was approximately one order of magnitude lower than that for oxygen-saturated water. Furthermore, in both cases, the F$1s$ signal completely disappeared and the intensity of the O$1s$ signal increased. These XPS results were interpreted in terms of the deprotonation of the pyrrole nitrogen occurring simultaneously with a rearrangement of the conjugation pattern of the aromatic polymer chains.

According to thermal degradation studies, the stability of PPY complexes is largely governed by dopants [127, 163]. In general, PPY/acceptor complexes, such as PPY/halide, PPY/perchlorate and PPY/halobenzoquinone complexes, exhibit a greater thermal stability than deprotonated PPY (consisting of 20–25 % =N– structure) and the fully reduced PPY° (consisting of 100 % –NH– structure). The temperatures for the onset of major weight loss in PPY/organic acceptor complexes such as the PPY/o-chloranil, PPY/p-chloranil and PPY/DDQ complexes, coincide approximately with the decomposition temperatures of the respective acceptors. The XPS data of the three PPY-halide complexes (PPY/iodide, PPY/bromide and PPY/chloride) suggested that two distinct weight loss processes occurred in these complexes [127]. Initially, the complex suffers a loss of the halide dopant, while with increasing temperature, the decomposition of the complex is shifted to a deprotonation process. The changes in the N$1s$ and halide core-level spectra are illustrated in Fig. 20 for the PPY/chloride complex. The Cl$2p$ core-level spectra of the pristine complex (Fig. 20(b)) and the complex after heating to 230 °C and 350 °C (Figs. 20(d) and 20(f), respectively) are best fitted with three spin-orbit split doublets (Cl$2p_{3/2}$ and Cl$2p_{1/2}$) having the Cl$2p_{3/2}$ peaks at about 197.1, 198.6 and 200.1 eV. These components are assigned to the Cl$^-$, Cl* and –Cl species, respectively (refer to Sect. 3.2.1). On the exposure to elevated temperature, both the Cl$^-$ and Cl* species are removed from the complex, with the initial rate of removal being somewhat faster for the latter than for the former. The initial loss of the chlorine species is not accompanied by a corresponding increase in the amount of the =N– component, as can be seen from the N$1s$ core-level spectrum of Fig. 20(c). However, at higher temperatures the chlorine species is lost (comparing Figs. 20(d) and 20(f)) with the concomitant appearance of deprotonated pyrrolylium nitrogens or the –N= structure (Fig. 20(e)).

In the case of PPY/perchlorate complex, a gradual decrease in the ClO$_4^-$/N ratio occurs when heated above room temperature [127, 164]. At sufficiently high temperatures (> 250 °C), the intensity of the Cl$2p$ core-level signal for the perchlorate species (BE in the 207 eV region) is greatly diminished, while that of

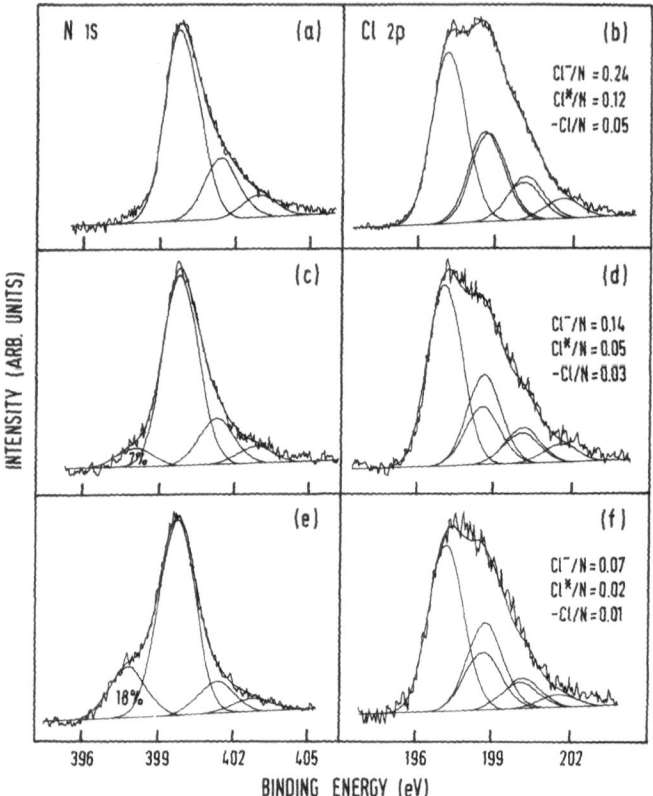

Fig. 20a–f. N$1s$ and Cl$2p$ core-level spectra of (**a**) and (**b**) pristine PPY/chloride complex (Cl/N = 0.41); (**c**) and (**d**) PPY/chloride complex heated to 230 °C in N$_2$ (Cl/N = 0.22); (**e**) and (**f**) PPY/chloride complex heated to 350 °C in N$_2$ (Cl/N = 0.10)

the chloride species (in the 200 eV region) is significantly enhanced. Partial redoping and halogenation of the polymer by these chloride species occur, but there remains a substantial amount of the imine component arising from the deprotonation of the pyrrolylium nitrogen. A similar decrease in the intensity of the ClO$_4^-$ peak and an increase in the peak intensity in the 200 eV region were observed when electrochemically prepared PPY/perchlorate films were stored in air for five months [165]. The same effect was also observed, although less pronounced, after storage for two weeks in ultra-high vacuum. The same study reported that ageing had a less pronounced effect in tosylate-doped PPY. In-situ spectroelectrochemical studies of doping and undoping of PPY with poly(ethylene oxide)–NaClO$_4$ solid electrolyte indicated a high degree of irreversibility of the electrochemistry of PPY with decomposition of ClO$_4^-$ anions and only a partial recovery of the N$1s$ spectrum with reoxidation of PPY film [44]. The propensity for halogens to form covalent bonds with the PPY backbone was also found during the simultaneous chemical polymerization and oxidation of pyrrole by halogens [166]. Of the three halogens (I_2, Br_2 and Cl_2) tested, the properties of the PPY/Br_2 complex showed the strongest dependence on the initial ratio of halogen to monomer. For this complex, the amount of bromine incorporated and the ratio of covalent bromide to ionic bromine increased upon increasing the initial bromine concentration in the reaction medium. The presence of substantial ring halogenation decreased the conductivity. The reactivity of the halogen acceptor affects its nature in the resulting complex. Thus, Cl_2 being the most reactive will be substituted at the β positions of the pyrrole rings even if a low reactant Cl_2 to pyrrole ratio is used, while I_2 being the least reactive will be present mainly as ionic iodide species.

In Sect. 3.1.2 we compared critically the chemical nature of the nitrogens corresponding to various intrinsic redox states of PPY and PAN, using XPS as the primary tool. The corresponding oxidation states of both polymers exhibit similar behavior towards protonation/deprotonation, oxidation/reduction, or CT interactions. However, the nitrogens of the two oxidized polymer complexes differ in thermal degradation behavior [47]. As mentioned before, the PPY/ chloride complex degrades via a deprotonation process, as suggested by the loss of the chloride species and also by the appearance of the deprotonated pyrro-lylium nitrogens or the =N– component in the N$1s$ core-level spectrum (Fig. 20). Figure 21 shows the N$1s$ and Cl$2p$ core-level spectra for a pristine emeraldine hydrochloride salt, EM/HCl (prepared by the oxidative polymeriz-ation of aniline by $(NH_4)_2S_2O_8$ in 1 mol l^{-1} HCl), and the sample after heat treatment at 300 °C in N_2. These spectra can be compared to those of Fig. 20. For the heat-treated EM/HCl complex, there is a loss of anion species, some of

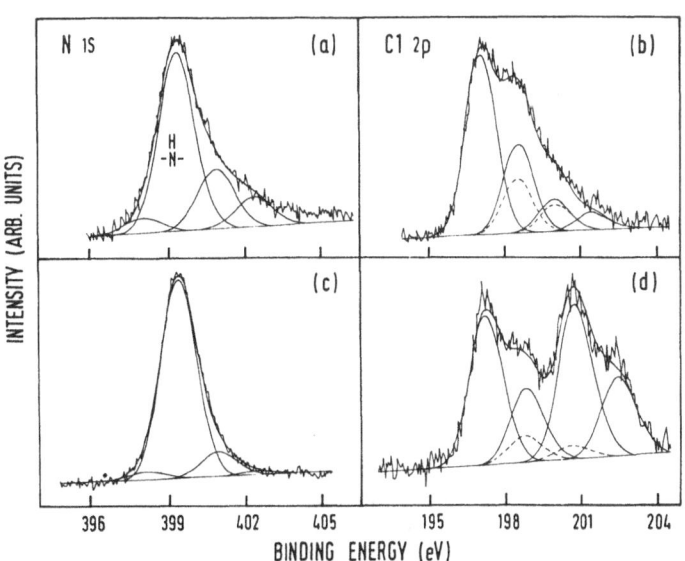

Fig. 21a–d. N$1s$ and Cl$2p$ core-level spectra of (**a**) and (**b**) pristine EM/HCl complex (Cl/N = 0.44); (**c**) and (**d**) EM/HCl complex heated to 300 °C in nitrogen (Cl/N = 0.14)

which in turn react with the polymer to increase the amount of covalently bonded chlorine substantially. Little deprotonation takes place, and the degraded sample exhibits a fairly symmetrical $N1s$ spectrum centered at about 399.3 eV and characteristic of the –NH– structure in PAN. Other thermal degradation studies of EM/HCl, either in vacuum [167] or N_2 [168], claimed that HCl began evolving around 200 °C, with the presence of HCl detected by either gas-phase IR spectroscopy or titration. If the volatile chlorine species were indeed HCl, and if little deprotonation accompanied the loss of chloride anions, the claim would imply that hydrogen is lost from the aromatic rings.

The conversion of chloride anions to covalently bonded chlorine in EM/HCl is detected even at 150 °C when air is present [169]. At this temperature, the process is retarded if air is absent [169]. In air, the conductivity of EM/HCl is temperature-dependent, being a bit thermally activated prior to a sharp decrease that occurs after about 60 °C [170]. The decrease in conductivity appears to be reversible if the sample is heated below 100 °C and then left standing in ambient air. With heat treatment at higher temperature, the conductivity will increase upon cooling but not to the level prior to heating. The recovery in conductivity is a slow process, and the water vapor in air appears to be an essential factor for it since only a minimal recovery of conductivity occurs when the heat-treated sample is cooled in a dessicator. It has been shown that water vapor increases the room temperature conductivities of both the base and the salt by a factor of about 2 [171]. However, the irreversible decrease in conductivity upon heating to above 100 °C would be consistent with the conversion of chlorine anions to covalent chlorine. The covalent bonding of chlorine to PAN also occurs during the oxidative polymerization of aniline by $(NH_4)_2S_2O_8$ in HCl if the acid concentration is high [36]. The XPS results show that for the EM/HCl polymers, the protonation level increases with the concentration of the protonic acid up to about 3 mol l^{-1}. At higher concentrations a considerable fraction of the chlorine incorporated is covalently bonded to the polymers even though a portion of the imine nitrogens remains unprotonated. For example, with 6 mol l^{-1} HCl, the Cl$^-$/N ratio is 0.12 while the –Cl/N ratio is 0.80.

The doping level (ClO$_4^-$/N ratio) of the PAN/perchlorate complex synthesized via the oxidative polymerization of aniline by $Cu(ClO_4)_2 \cdot 6H_2O$ in acetonitrile [172] does not change significantly even after 24 h at 150 °C in air [164]. This is in marked contrast to the Cl$^-$ anions in EM/HCl. However, the conductivity of the perchlorate complex does not exhibit improved stability over that of the EM/HCl complex. Thus the former must undergo structural changes that are not manifested in the surface stoichiometries as determined by XPS. From a thermogravimetric analysis of the PAN/perchlorate complex, a rapid weight loss step, attributable to the decomposition of the ClO$_4^-$ anions, was found to occur at about 250 °C. Similar behavior was observed with PPY/perchlorate complex [164], but this complex was more susceptible to deprotonation. The PAN/perchlorate complex can also be prepared by oxidative doping of LM by $Cu(ClO_4)_2 \cdot 6H_2O$. When acetonitrile is used as the reaction medium, its $N1s$ core-level spectrum is similar to that of the complex obtained from the simultaneous polymerization and oxidation of aniline by $Cu(ClO_4)_2 \cdot 6H_2O$ [173]. In these complexes, no significant imine structures are formed. In contrast, oxidation of some of the amine units of LM to the imine structure by $Cu(ClO_4)_2 \cdot 6H_2O$ can occur if THF, MeOH or H_2O is used. In these solvents, the doping level is substantially lower than that obtained with the use of acetonitrile. Furthermore, when the PAN/perchlorate complex is exposed to protic solvents such as methanol or H_2O, a fraction of the ClO_4^- anions is removed [173], and this is accompanied by a decrease in the proportion of positively charged nitrogens (and thus conductivity) and a commensurate increase in the amine as well as imine nitrogens. The increase in the amine units implies that a large fraction of ClO_4^- anions is removed from the complex with no simultaneous breaking of the N–H bond of the nitrogenonium structure. This phenomenon differs from the deprotonation that occurs when this complex is treated with NaOH. The treatment of EM/HCl with organic solvents would also affect its conductivity owing to the loss of chloride anions and/or the conversion of some of the anions to covalently bonded chlorine [170].

Although PAN salts have so far been considered insoluble in most of common solvents, a recent report has revealed that they can be dissolved completely in concentrated H_2SO_4 [174]. When protonic acid dopants of large molecular size such as toluene-p-sulfonic acid were used in the chemical synthesis of PAN, a salt soluble in common organic solvents was obtained [175]. The PAN/perchlorate complex from the simultaneous oxidation and polymerization of aniline by $Cu(ClO_4)_2 \cdot 6H_2O$ was also found to be soluble in DMSO. NMP was shown to be a good solvent for casting free-standing EM base film [23]. The PAN/perchlorate complex can also be dissolved in this solvent to give a thick green solution from which film can be cast. However, the addition of excess NMP will lead to deprotonation of the complex, as evidenced by the UV-visible absorption spectrum of the blue solution obtained. Furthermore, while the addition of $HClO_4$ to an NMP solution of the EM base will cause protonation in solution, the protonated species are only metastable and slowly undergoes deprotonation [133]. Due to these difficulties, PAN salt films can be more easily prepared by the protonation of EM base films with dilute acids. A study of the structure of PAN film after progressive and repeated reprotonation/deprotonation has shown that structural modifications occur after one reprotonation/deprotonation cycle [176]. The $N1s$ spectra show that when the base film is reprotonated to less than 20% by dilute $HClO_4$ (Fig. 22(a)) and then deprotonated (Fig. 22(b)), the composition of the resulting base remains similar to that of the pristine base. However, if the doping level is increased to 50% during reprotonation (Fig. 22(c)), the resulting base film will be in an oxidation state, characterized by an imine/amine ratio significantly higher than that of the pristine base (Fig. 22(d)). Moreover, the film can be easily reprotonated to about 85% (as determined from N^+/N and ClO^-_4/N ratios) with 3 mol l^{-1} $HClO_4$ (Fig. 22(e)), whereas in the corresponding case of EM base powder, the protonation level is $\sim 50\%$. This would imply the protonation of a large fraction of the

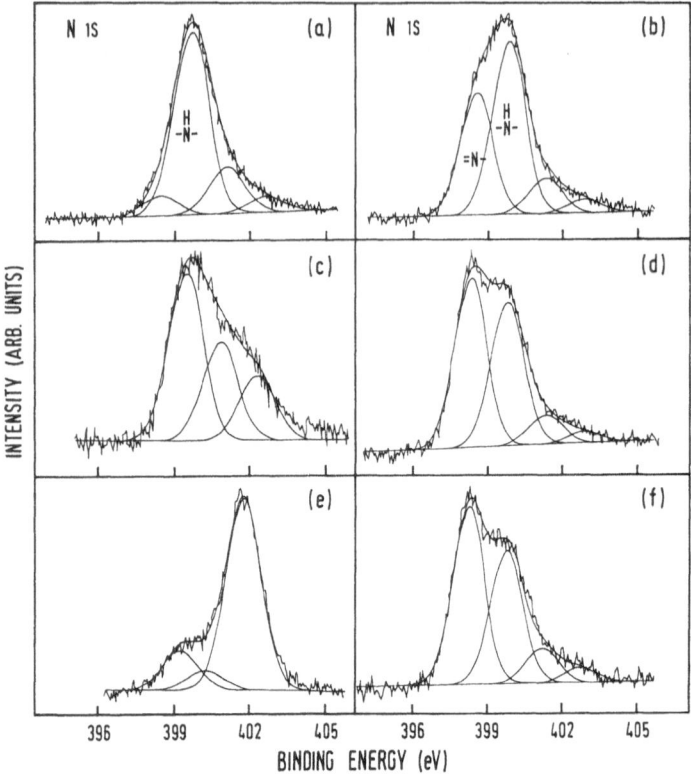

Fig. 22a–f. N*1s* core-level spectra of EM films: (**a**) reprotonated with 0.0 2 M HClO$_4$ (Sample 1); (**c**) reprotonated with 1.0 mol l^{-1} HClO$_4$ and rinsed with 0.1 mol l^{-1} HClO$_3$ (Sample 2); (**e**) reprotonated with 3.0 mol l^{-1} HClO$_3$ (Sample 3); (**b**), (**d**) and (**f**) Samples 1, 2 and 3 after treatment with 0.5 mol l^{-1} NaOH

amine units of the film in addition to all the imine units. The N*1s* core-level spectra of the base films also indicate an increase in the BE separation between the imine and amine components peaks, which amounts to 1.20 eV for the pristine base powder and film and increases to about 1.45–1.50 eV for the base films obtained from the highly protonated salt films. (Figs. 22(d) and 22(f)). When HCl is used as the protonating acid for the films, the bases after reprotonation/deprotonation also show differences from the pristine base, albeit not to the same extent as when HClO$_4$ is used. These structural modifications are also manifested in IR absorption spectra and drastically reduce the solubility of the base films. However, no significant changes are observed in the subsequent reprotonation process, but only in the third reprotonation cycle, the chemical state of the anions shows an apparent change. For the EM powder samples, the reprotonation/deprotonation process appears to be reversible over the three cycles tested, causing no apparent changes in the structure (as indicated by XPS and IR spectra) of polyaniline.

It should be noted that although 75% of the nitrogens in EM can be positively charged via the protonation of the amine as well as imine units, the protonation of a 75% intrinsically oxidized NA base by $1 \, mol \, l^{-1}$ $HClO_4$ causes the imine nitrogens to disappear completely, but the protonation level remains at 50% [177]. This is probably due to the hydrolysis of some quinonoid units in the aqueous medium [23]. The degradation of PAN via the formation of p-benzoquinone has been observed in electrochemical studies [178, 179]. The voltammetric response of PAN film persisted during the repetition of potential cycling between -0.2 V and 0.6 V relative to the standard calomel electrode in $1 \, mol \, l^{-1}$ HCl, while the film properties degraded when the anodic limit of the potential cycling exceeded 0.7 V.

A comparative study on electrochemically synthesized PPY and PBiT perchlorate complexes has established, from the XPS spectral linewidths, that the latter is considerably more ordered than the former [15]. The stability of PTHs depends strongly on counter-ions. For example, the conductivity of electrochemically synthesized PTH/BF_4 and PTH/ClO_4 decreased markedly when heated in air to above $70 \, °C$ [117, 118], while electrochemically synthesized PTH doped with $FeCl_3$ was thermally more stable than the former two complexes [180]. The thermal degradation of PBiT/perchlorate complexes also proceeds via the decomposition of the ClO_4^- anions and the partial conversion to chloride species, but this process occurs at a lower temperature for the PBiT complex than for the corresponding PPY complex [164]. The ClO_4^- doped PBiT suffers a loss of anions and a concomitant decrease in conductivity when treated with methanol [181]. Tourillon and Garnier [141] claimed that the doping level and conductivity of electrochemically synthesized PMeT with $CF_3SO_3^-$ counterions remained unchanged during the 8-month storage in air. The long-term stability was also observed with the undoped PMeT. No oxygen or H_2O was detected in this sample during the 8-month storage in ambient air, in contrast to PPY which has been shown to be very sensitive to oxygen. The stabilities of PPY and PMeT under electrochemical treatment were compared. Their current-voltage (I–V) and XPS characteristics were recorded before and after 20 cycles of polarization between their oxidized and reduced neutral states in the $N(Bu)_4PF_6/CH_3CN$ medium. For PPY the oxidation and reduction peaks were not well defined and the XPS spectra were greatly modified: the peak due to P was no longer detected, the intensity of the F$1s$ peak decreased but the O$1s$ peak increased. These indicate the replacement of the PF_6^- dopant by O_2 during the electrochemical treatment. In contrast, PMeT maintained good electrochemical stability, because the I–V curve and XPS spectra underwent no change after cyclic polarizations. Doping-undoping tests were carried out for PTH and PMeT in an aqueous electrolytic medium with salts such as K_2PtCl_6 over 500 cycles [182]. The electrochemical and XPS data recorded after the second and 500th cycles showed only minor differences. However, in another study [104] where the doping-undoping of PTH was carried out in the range of -2 to $+4$ V in a $LiBF_4/CH_3CN$ solution for 1.35×10^5 cycles, the XPS data indicated degradation of the film. In the degraded film, no B$1s$ signal was observed but an $F1s$ signal was present, and the intensity of the high BE component of the $C1s$ peak at 289.6 eV increased. It was postulated that F^- formed from the decomposition of BF_4^- reacts with C. Depth profiling indicated that the degradation penetrated down to the bulk of the polymer film.

The presence of alkyl chains on the polythiophene backbone renders the polymer soluble and hence improves its processability. The solvatochromism in solution and the thermochromism in the solid state of poly(3-alkylthiophene)s have been interpreted in terms of conformational changes (see Sect. 3.3.1). The rapid thermal undoping of doped PHeT has been postulated to be due to similar conformational changes on the basis of the reversible thermochromic effect [183–185]. The degradation rate is dependent on the surrounding atmosphere. The $Fe2p$ core-level spectra of $FeCl_3$-doped PHeT before and after heat treatment at 195 °C for 2 h showed a shift of the characteristic peaks to lower BE, which indicates that the Fe species changes from $FeCl_3$ or $FeCl_4^-$ to $FeCl_2$ [186]. This indication was confirmed by the disappearance of the $FeCl_4^-$ peaks in the optical spectra (3.4, 3.9 and 5.1 eV) of the doped polymer film upon heating. A comparison of the degradation rate of the PHeT with that of PMeT suggests that the long alkyl chain plays a role in the more rapid thermal undoping of the former. It has been postulated [17] that the oxidation potential of PHeT increases at elevated temperature owing to an increased number of twists (disruptions of planarity) on the polymer chain. In solid material, the undoping is irreversible, whereas the optical changes induced in the doped solution are reversible. This difference is believed to be due to the products resulting from the undoping of the solid forming stable complexes, while in solution, the products are solvated and may be able to redope the chain upon cooling.

An attempt has been made to enhance the stability of doped PHeT by using an anion which is not easily oxidizable and also by hindering the geometrical distortion of the polymer chain [187]. Sodium dodecyl benzene sulfonate was chosen as the supporting electrolyte in methanol for the doping of PHeT. The presence of sulfonate species in the polymer was clearly documented by the XPS spectra, but surface segregation of the dopant on the outermost layer of the polymer film may not be ruled out. Loss of the anion, or redistribution of the anions was also seen in the spectra after thermal treatment of 120 °C for 2 h in vaccum. Thus, the alkyl-sulfonate doped PHeT does not resolve the problem concerning thermal undoping. The conductivity of this material is lower by several orders of magnitude than that of $FeCl_3$-doped poly(alkylthiophene)s.

4 Some Future Directions

It is fair to say that during the past one and a half decades, the majority of research efforts in electroactive polymers has aimed at enhancing our fundamental knowledge on the structure-property relationship in this class of materials. Undoubtedly, XPS has played an important role in this endeavor. The next stage of development in electroactive polymers will be in the applications of these materials to electronic devices such as conductors, sensors and active electrodes. As in the case of all other polymers, this step would have to go with materials processing and modification. In such directions, XPS will continue to play a useful role as a versatile surface analysis technique. In fact, XPS has already proven to be unique in quantitative tracking of the changes that occur in the intrinsic structure of PAN during the powder-to-film processing by solution casting [176].

The applications of electroactive polymers to electrodes and sensors may require substantial material modification, in particular, surface modification. For conventional polymers, such as polyethylene, poly(ethylene terephthalate), poly(vinyl chloride), nylons and polypropylene, surface modification using biocompatible materials for biochemical and biomedical applications has been well established [188, 189]. It has also been demonstrated that surface modification can be performed through, for example, graft copolymerization under mild conditions when the surface is pretreated with high energy radiation, glow discharge, corona discharge, ozone exposure or UV irradiation. Furthermore, protein and enzyme immobilizations on the surface modified polymer substrates have also been reported [190]. In view of the reactive nature of most electroactive polymer surfaces, surface modification and functionalization via molecular design may eventually prove to be a fruitful area for future research. This is because electroactive polymer substrates may have an added advantage over conventional polymer substrates. XPS and related surface analysis techniques will be essential in this undertaking [191].

The typical sampling depth of 5 to 10 nm makes the XPS technique inadequate in some cases for use in the study of reactive conjugated polymer surfaces. The surface sensitivity can be increased somewhat, at least for smooth surfaces, by the use of angular variation measurements, in which the electron take-off angle (with respect to the sample surface) is reduced. Alternatively, the XPS technique can be supplemented by more surface sensitive techniques, such as secondary ion mass spectroscopy (SIMS). In this spectroscopic technique, the polymer surface is sputtered by a noble gas ion beam, such as Ar^+ or Xe^+ of ≤ 4 KeV and $\sim lnA/cm^2$ current density [192]. The small percentages of the positively and negatively charged fragments are extracted into a mass spectrometer (typically of the quadrupole type) and provide, in sequential experiments, the positive and negative secondary ion mass spectra, respectively. The following example illustrates the use of combined XPS and SIMS techniques in probing the intrinsic structures of the highly reactive $(CH)_x$ surface.

As mentioned in Sect. 3.1.1, the as-prepared $(CH)_x$ film surface is usually oxidized to some extent. Figure 23(a) and Fig. 23(b) show, respectively, the negative static SIMS spectra of a high density cis-$(CH)_x$ film before and after sputtering by Ar^+ beam [193]. The negative ion spectrum for the as-received film is dominated in the low mass region by signals due to C_2H^- (25 amu), C_2^-

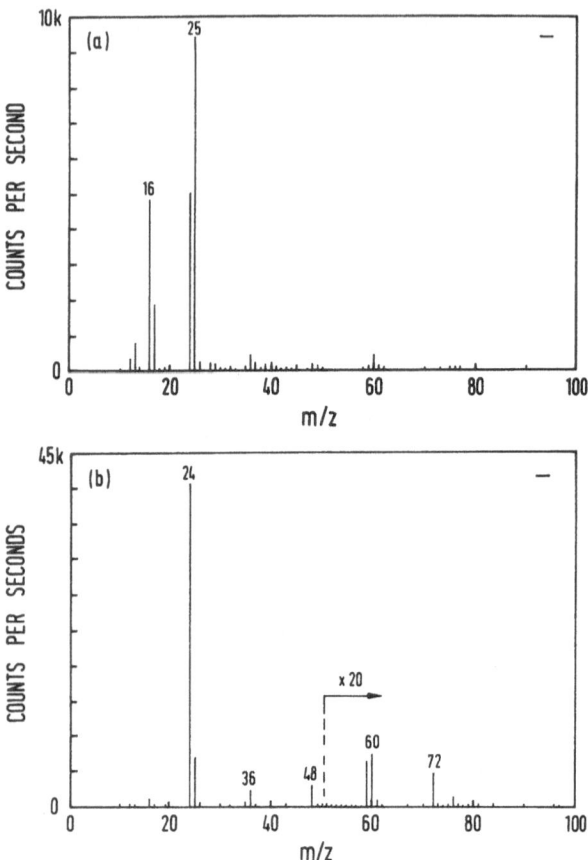

Fig. 23a, b. Negative SIMS spectra of (**a**) as-received and (**b**) sputtered $(CH)_x$ film

(24 amu) and O^- (16 amu), the largest being the C_2H^- signal. After light Ar^+ ion sputtering, the C_2^- species predominates and other C_n^- (n = 1–8) species are also clearly discernible. The C_2H^- and O^- signals have decreased significantly. Comparison of the $C_2^- : C_2H^-$ ratio of the lightly sputtered $(CH)_x$ surface with those of the surfaces of other saturated and conjugated polymers indicates that a $C_2^- : C_2H^-$ ratio greater than one is characteristic of a conjugated main chain. Thus, due to its highly reactive nature, the spectrum of the as-received $(CH)_x$ surface exhibits features typical of those of the surfaces of a saturated and a partially oxidized polymer, simultaneously. After light spluttering, the C1s core-level spectrum becomes rather symmetrical, which is consistent with the presence of predominantly single carbon species expected for $(CH)_x$. It should be noted, however, that effects of prolonged sputtering on the polymer surface are not yet well understood.

Finally, we have to remark that the conventional XPS is not capable of elucidating the detailed surface morphology of polymer films. This shortcoming can be greatly alleviated by coupling it with the scanning tunneling microscopic (STM) technique. A number of recent STM studies on PPY [194] and PAN [195], as well as a combined XPS and STM study on PAN [196], have provided very fruitful results. Furthermore, the imaging XPS technique is coming fast into sight [197].

5 Summary

In this review, some of the electroactive polymers most commonly studied during the past one and a half decades have been selected to illustrate the type and level of information obtainable from XPS core-level spectra. It concerns (a) the intrinsic structure, (b) the CT interaction, and (c) the stability and degradation behavior. The review is meant to be comprehensive, although emphasis has been placed on some specific issues related to these three basic physicochemical properties. For example, the chemical nature of the nitrogens in PPY and PAN has been critically compared on the basis of XPS data. Some of the major discrepancies in the XPS literature of electroactive polymers have also been examined. In most cases, preference has been given to results for which proper justification and careful comparison with available data are possible. Finally, some future trends in the application of XPS and other more surface sensitive techniques to the study of highly reactive conjugated polymer surfaces have been mentioned.

Acknowledgement: We are deeply indebted to Professor H.H. Huang, Deputy Vice-Chancellor of the National University of Singapore, for spearheading the establishment of the Surface Science Laboratory. We are grateful to Professor H. Fujita for giving us the opportunity to write this review and for this critical and helpful comments on the manuscript. We also wish to thank Professor Der-Jang Liaw of the National Taiwan Institute of Technology for his constant encouragement.

6 References

1. See for example: Skotheim T (ed) (1986) Handbook of conducting polymers, vols I and II. Marcel Dekker, New York
2. See for example: Muilenberg GE (ed) (1977) Handbook of X-ray photoelectron spectroscopy. Perkin Elmer, New York.
3. Salaneck WR (1986) In: Skotheim T (ed) Handbook of conducting polymers, vol II. Marcel Dekker, New York, p 1337
4. Shirakawa H, Ikeda S (1971) Polym J 2: 231
5. Chiang CK, Fincher Jr CR, Park YW, Heeger AJ, Shirakawa H, Louis EJ, Gau SC, MacDiarmid AG (1977) Phys Rev Lett 93: 1098
6. Seah MP, Briggs D (1983) In: Briggs D, Seah MP (eds) Practical surface analysis by Auger and X-ray photoelectron spectroscopy. John Wiley, Chichester, p 6

7. Seah MP, Dench WA (1979) Surf Interf Anal 1: 2
8. Riviere JC (1983) In: Briggs D, Seah MP (eds) Practical surface analysis by Auger and X-ray photoelectron spectroscopy. John Wiley, Chichester, p 17
9. Siegbahn K, Nordling C, Fahlman A, Nordberg R, Hamrin K, Hodman J, Johansson G, Bergmark T, Karlsson S, Lindgren I, Linderg B (1967) ESCA: atoms, molecules and solid state structure studied by means of electron spectroscopy. Almquist and Wiksells, Uppsala
10. Clark DT (1981) In: Dwight DW, Fabish TJ (eds) Photons, electrons and ion probes of polymer structure and properties. Am Chem Soc Washington DC, p 255 (ACS Symposium Series 162)
11. Dilks A (1981) In: Brundle CR, Baker AD (eds) Electron spectroscopy – theory, techniques and applications, vol 4. Academic, London, p 278
12. Briggs D (1983) In: Briggs D, Seah MP (eds) Practical surface analysis by Auger and X-ray photoelectron spectroscopy. John Wiley, New York, p 359
13. Dwight DW, McGrath JE, Wightman JP (1978) J Appl Polym Sci Symp 34: 35
14. Briggs D, Riviere JC (1983) In: Briggs D, Seah MP (eds) Practical surface analysis by Auger and X-ray photoelectron spectroscopy. John Wiley, New York, p 114
15. Pfluger P, Street GB (1984) J Chem Phys 80: 545
16. Salaneck WR, Thomas HR, Bigelow RW, Duke CB, Plummer EW, Heeger AJ, MacDiarmid AG (1980) J Chem Phys 72: 3674
17. Munstedt H (1988) Polymer 29: 296
18. Kang ET, Neoh KG, Tan KL, Tan BTG (1990) J Polym Sci Polym Phys Ed 27: 2061
19. Kang ET, Neoh KG, Tan KL (1991) J Polym Sci Polym Phys Ed 29: 669
20. Masuda T, Higashimura T (1987) In: Okamura S (ed) Adv in Polym Sci 81. Springer, Berlin Heidelberg New York, p 121
21. Diaz AF, Kanazawa KK (1983) In: Miller JS (ed) Extended Linear Chain Compounds, vol 3. Plenum, New York, p 417
22. Ray A, Asturies GE, Kershner DL, Richter AF, MacDiarmid AG, Epstein AJ (1989) Synth Met 29: E141
23. Angelopoulos M, Asturias GE, Ermer SP, Ray A, Scherr EM, MacDiarmid AG, Akhtar M, Kiss Z, Epstein AJ (1988) Mol Cryst Liq Cryst 160: 151
24. Cao Y, Smith P, Heeger AJ (1989) Synth Met 31: 263
25. Phillips SD, Yu G, Gao Y, Heeger AJ (1989) Phys Rev B 39: 10702
26. Jozefowicz ME, Laversanne R, Javadi HHS, Epstein AJ, Pouget JP, Tang X, MacDiarmid AG (1989) Phys Rev B 39: 12958
27. Genies EM, Boyle A, Lapkowski M, Tsintavis C (1990) Synth Met 36: 139
28. Snauwaert P, Lazzaroni R, Riga J and Verbist JJ (1986) Synth Met 16: 245
29. Snauwaert P, Lazzaroni R, Riga J and Verbist JJ (1987) Synth Met 18: 335
30. Kang ET, Neoh KG, Khor SH, Tan KL, Tan BTG (1989) J Chem Soc Chem Commun 1989: 695
31. Tan KL, Tan BTG, Kang ET, Neoh KG (1989) Phys Rev B 39: 8070
32. Kang ET, Neoh KG, Tan KL, Kuok MH, Tan BTG (1990) Mol Cryst Liq Cryst 178: 219
33. Snauwaert P, Lazzaroni R, Riga J, Verbist JJ, Gonbeau D (1990) J Chem Phys 92: 2187
34. Kumar SN, Gaillard F, Bouyssoux G, Sartre A (1990) Synth Met 36: 11
35. Cao Y (1990) Synth Met 35: 319
36. Tan KL, Tan BTG, Khor SH, Neoh KG, Kang ET (1991) J Chem Phys Solids 52: 673
37. Kang ET, Neoh KG, Tan KL (1989) Polym J 21: 873
38. Kang ET, Neoh KG, Tan KL, Tan BTG (1990) Synth Met 35: 345
39. Nakajima T, Harada M, Osawa R, Kawagoe T, Furukawa Y (1989) Macromolecules 22: 2644
40. Monkman AP, Bloor D, Stevens GC, Stevens JCH (1989) In: Kuzmany H, Mehring M, Roth S (eds) Springer Ser in Solid-State Sci 91. Springer, Berlin Heidelberg New York, p 295
41. Monkman AP, Stevens GC, Bloor D (1991) J Phys D: Appl Phys 24: 738
42. Inganäs O, Erlandsson R, Nylander C, Lundström I (1984) J Chem Phys Solids 45: 427
43. Kang ET, Neoh KG, Ong YK, Tan KL, Tan BTG (1990) Synth Met 39: 69
44. Skotheim TA, Florit MI, Melo A, O'Grady WE (1984) Phys Rev B 30: 4846
45. Zeller MV, Hahn SJ (1988) Surf Interf Anal 11: 327
46. Gustafsson G, Lunström I, Liedberg B, Wu CR, Inganäs O, Wennerstrom O (1989) Synth Met 31: 163
47. Tan KL, Tan BTG, Kang ET. Neoh KG (1991) J Chem Phys 94: 5382
48. Kang ET, Neoh KG, Ong YK, Tan KL, Tan BTG (1991) Macromolecules 24: 2822
49. Yoshino K, Nakajima S, Sugimoto R (1987) Jpn J Appl Phys 26: L1038

50. Hotta S, Rughooputh SDV, Heeger AJ (1987) Synth Met 22: 79
51. Mastragostino M, Marinangli AM, Corradini A, Giacobbe S (1989) Synth Met 28: C501
52. Salaneck WR, Inganäs O, Nilsson JO, Österholm, Themans B, Bredas JL (1989) Synth Met 28: C451
53. Clarke TC, Kanazawa KK, Lee VY, Rabolt JF, Reynolds JR, Street GB (1982) J Polym Sci Polym Phys Ed 20: 117
54. Shacklette LW, Elsenbaumer RL, Chance RR, Eckhardt H, Frommer JE, Baughman RH (1981) J Chem Phys 75: 1919
55. Riga J, Snauwaert PH, DePryck A, Lazzaroni R, Boutique JP, Verbist JJ, Bredas JI, Andre JM, Taliani C (1987) Synth Met 21: 223
56. Tourillon G, Jugnet Y (1988) J Chem Phys 89: 1905
57. Keane MP, Svensson S, Naves de Brito A, Correia N, Lunell S, Sjogren B, Inganas O, Salaneck (1990) J Chem Phys 93: 6357
58. Gourley KD, Lillya CP, Reynolds JR, Chien JCW (1984) Macromolecules 17: 1025
59. Ohnishi T, Noguchi T, Nakano T, Hirooka M, Murase I (1991) Synth Met 41: 309
60. Obrzut MJ, Karasz FE (1989) Macromolecules 22: 458
61. Nguyen TP, Ettaik H, Lefrant S, Leising G, Stelzer F (1990) Synth Met 38: 69
62. Ng KT, Hercules DM (1975) J Am Chem Soc 97: 4168
63. Hsu SL, Signorelli AJ, Pez GP, Baughman RH (1978) J Chem Phys 69: 106
64. Ikemoto I, Sakairi M, Tsutsumi T, Kuroda H, Harada I, Tasumi M, Shirakawa H, Ikeda S (1979) Chem Lett (Jpn) 1979: 1189
65. Petit MA, Soum AH, Leclerc M, Prudhomme RE (1987) J Polym Sci Polym Phys Ed 25: 423
66. Ikemoto I, Cao Y, Yamada M, Kuroda H, Harada I, Sirakawa H, Ikeda S (1982) Bull Chem Soc Jpn 55: 721
67. Asakura K, Ikemoto I, Kuroda H, Kobayashi T, Shirakawa H (1985) Bull Chem Soc Jpn 58: 2113
68. Salaneck WR, Thomas HR, Duke CB, Plummer EW, Heeger AJ, MacDiarmid AG (1980) Synth Met 1: 133
69. Clarke TC, Geiss RN, Gill WD, Grant PM, Mackling JW, Morawitz H, Rabolt JF, Street GB, Sayers D (1979) J Chem Soc Chem Commun 1979: 332
70. Brant P, Moran MJ, Weber DC (1989) Chem Phys Lett 76: 529
71. Wu HM, Chen SA (1987) Polym Commun 28: 75
72. Ikemoto I, Ichihara T, Egawa C, Kikuchi K, Kuroda H, Furukawa Y, Harada I, Shirakawa H (1985) Bull Chem Soc Jpn 58: 747
73. Ehrlich P, Anderson W (1986) In: Skotheim T (ed) Handbook of conducting polymers, vol I. Marcel Dekker, New York, p 441
74. Kang ET, Ehrlich P, Bhatt AP, Anderson W (1984) Macromolecules 17: 1020
75. Kang ET, Neoh KG, Masuda T, Higashimura T, Yamamoto M (1989) Polymer 30: 1328
76. Russo MV, Polzonetti G, Furlani A (1991) Synth Met 39: 291
77. Salaneck WR, Erlandsson R, Prejza J, Lundström I, Inganäs O (1983) Synth Met 5: 125
78. Munro HS, Parker D, Eaves JG (1987) In: Kuzmany H, Mehring M, Roth S (eds) Springer Ser in Solid-State Sci 76. Springer, Berlin Heidelberg New York, p 257
79. Eaves JG, Munro HS, Parker D (1987) Polym Commun 28: 38
80. Kang ET, Ti HC, Neoh KG, Tan TC (1988) Polym J 20: 399
81. Tan KL, Tan BTG, Kang ET, Neoh KG (1989) J Materials Sci 25: 805
82. Kang ET, Neoh KG, Ong YK, Tan KL, Tan BTG (1991) Polymer 32: 1354
83. Dannetunn P, Lazzaroni R, Salaneck WR, Scherr E, Sun Y, MacDiarmid AG (1991) Synth Met 41: 645
84. Mirrezaei SR, Munro HS, Parker D (1988) Synth Met 26: 169
85. Ribo JM, Dicko A, Tura JM, Bloor D (1991) Polymer 32: 728
86. Epstein AJ, MacDiarmid AG (1988) J Mol Elect 4: 161
87. Bredas JL, Themans B, Andre JM, Chance RR, Sibley R (1984) Synth Met 9: 265
88. Machida S, Miyata S, Techagumpuch T (1989) Synth Met 31: 311
89. Pron A, Kucharski Z, Budrowski C, Zagorska M, Krichene S, Suwalski J, Dehe G, Lefrant S (1985) J Chem Phys 83: 5923
90. Inoue MB, Nebensny KW, Fernando Q (1990) Synth Met 38: 205
91. Lyons AM, Vasile MJ, Pearce EM, Waszezak JV (1988) Macromolecules 21: 3125
92. Baughman RH, Wolf JF, Eckhardt H, Shacklette LW (1988) Synth Met 25: 121
93. Kang ET, Neoh KG, Khor SH, Tan KL, Tan BTG (1990) Polymer 31: 202

94. Kang ET, Neoh KG, Tan TC, Khor SH, Tan KL, Tan BTG (1990) Macromolecules 23: 2918
95. Tanaka K, Wang SL, Yamabe T (1990) Synth Met 36: 129
96. Kang ET, Neoh KG, Tan KL, Tan BTG (1992) Synth Met 46: 227
97. Ferraro JR, Williams JM (1987) Introduction to synthetic electrical conductors. Academic, New York p 22
98. Khor SH, Neoh KG, Kang ET (1990) J Appl Polym Sci 40: 2015
99. Yue J, Epstein AJ (1991) Macromolecules 24: 4441
100. Kang ET, Neoh KG, Woo YL, Tan KL (1991) Polym Commun 32: 412
101. Hotta S, Shimotsuma W, Taketani M, Kohiki S (1985) Synth Met 11: 139
102. Wu CR, Nilsson JO, Inganäs O, Salaneck WR, Österholm JE, Bredas JL (1987) Synth Met 21: 197
103. Fink J Nucker N, Scheerer B, Neugebauer H (1987) Synth Met 18: 163
104. Takenaka Y, Koike Tomoyuki, Oka T, Tanahashi M (1987) Synth Met 18: 207
105. Nilsson JO, Inganäs O (1989) Synth Met 31: 359
106. Kang ET, Neoh KG, Tan KL (1991) Phys Rev B 44: 10461
107. Morea G, Sabbatini L, Zambonin PG, Swift AJ, West RB, Vickerman JC (1991) Macromolecules 24: 3630
108. Tsukamoto J, Matsumura K (1985) Jpn J Appl Phys 24: 974
109. Tsukamoto J, Fukuda S, Tanaka K, Yamabe T (1987) Synth Met 17: 673
110. Simizu H, Tanabe Y, Kanetsuna H (1986) Polym J 18: 367
111. Philips GM, Untereker DF (1980) In: Owners BB, Nargalli N (eds) Proc Electrochem Society – Power sources biomedical inplantable applied ambient temperature Li batteries, Electrochem Soc. Pennington, NJ, 80-4: 195
112. Audenaert M, Gusman G, Mehbod M, Deltour R, Noirhomme B, Donckt EV (1969) Solid State Commun 30: 797
113. Tan KL, Tan BTG, Kang ET. Neoh KG (1990) J Mol Elect 6: 5
114. Tan KL, Tan BTG, Kang ET, Neoh KG (1989) J Appl Phys 66: 5868
115. Pearson JM, Stolka M (1981) Polyn(N-vinylcarbazole), Gorden and Breach, New York
116. Hino S (1987) Synth Met 18: 253
117. Dubois JE, Desbene A, Lacaze PC (1982) J Electroanal Chem 132: 177
118. Billingham NC, Calvert PD, Foot PJS, Mohammad F (1987) Poly Deg Stab 19: 323
119. Deits W, Cukor P, Rubner M, Jopson H (1982) Synth Met 4: 199
120. Pochan JM, Gibson HW, Bailey FC (1980) J Polym Sci Polym Lett Ed 18: 447
121. Pochan JM, Gibson HW, Bailey FC, Pochan DF (1980) Polymer 21: 250
122. Masuda T, Tang BZ, Higashimura T, Yamaoka H (1985) Macromolecules 18: 2369
123. Neoh KG, Kang ET, Tan KL (1989) Thermochimica Acta 146: 251
124. Neoh KG, Kang ET, Tan KL (1990) Polym Deg Stab 29: 279
125. Neoh KG, Kang ET, Tan KL (1989) Polym Deg Stab 26: 21
126. Neoh KG, Kang ET, Tan KL (1991) Polymer 32: 226
127. Kang ET, Neoh KG, Ong YK, Tan KL, Tan BTG (1991) Thermochimica Acta 181: 57
128. Diaz AF, Castillo JI, Logan JA, Lee WY (1981) J Electroanal Chem 129: 115
129. Qian R, Qiu J (1987) Polym J 18: 157
130. Pfluger P, Krounbi M, Street GB, Weiser G (1983) J Chem Phys 78: 3212
131. Neoh KG, Kang ET, Tan KL (1990) Thermochimica Acta 171: 279
132. Tang J, Jing X, Wang B, Wang F (1988) Synth Met 24: 231
133. Neoh KG, Kang ET, Tan KL (1992) Polymer (In Press)
134. Masuda H, Tanaka S, Kaeriyama K (1989) Synth Met 33: 365
135. Rühe J, Esquerra TA, Wegner G (1989) Synth Met 28: C177
136. Sato M, Tanaka S, Kaeriyama K (1986) J Chem Soc Chem Commun 1986: 876
137. Jen KY, Miller GG, Elsenbaumer RL (1986) J Chem Soc Chem Commun 1986: 1346
138. Inganäs O, Salaneck WR, Österholm JE, Laakso J (1988) Synth Met 22: 395
139. Salaneck WR, Inganäs O, Thémans B, Nilsson JO, Sjögren B, Österholm JE, Brédas JL, Svensson S (1988) J Chem Phys 89: 4613
140. Inganäs O, Gustafsson G, Salaneck WR, Österholm JE, Laakso J (1989) Synth Met 28: C377
141. Tourillon G, Garnier F (1983) J Electrochem Soc 130: 2042
142. Kaul A, Udipi K (1989) Macromolecules 22: 1201
143. Chiang CK, Park YW, Heeger AJ, Shirakawa H, Louis EJ, MacDiarmid AG (1978) J Chem Phys 69: 5098
144. Shirakawa H, Louis EJ, MacDiarmid AG, Chiang CK, Heeger AJ (1978) J Chem Soc Chem Commun 1978: 578

145. Chiang CK, Gau SC, Fincher Jr CR, Park YW, MacDiarmid AG, Heeger AJ (1978) Appl Phys Lett 33: 18
146. Inoue T, Österholm JE, Yasuda HK, Levensen LL (1980) Appl Phys Lett 36: 101
147. Harada I, Tasumi M, Shirakawa H, Ikeda S (1978) Chem Lett (Jpn) 1979: 1411
148. Rolland M, Lefrant S, Aldissi M, Bernier P, Rzepka E, Shue F (1981) J Electron Mat 10: 619
149. Österholm JE, Yasuda HK, Levenson LL (1983) J Appl Polym Sci 28: 1265
150. Pochan JM (1986) In: Skotheim T (ed) Handbook of conducting polymers, vol II. Marcel Dekker, New York, p 1383
151. Huq R, Farrington GC (1984) J Electrochem Soc 131: 819
152. Druy MA, Rubner MF, Walsh SP (1986) Synth Met 13: 207
153. Guiseppi-Elie A, Wnek GE (1990) J Phys Chem 94: 3192
154. Chen YC, Ni FL, Shiramatsu T, Tseng JS (1984) Polym Commun 25: 315
155. Lee MS, Tseng JS, Chen YC (1989) Synth Met 31: 191
156. Ahlgren G, Krische B (1984) J Chem Soc Chem Commun 1984: 703
157. McNeill R, Sindak R, Wardlaw JH, Weiss (1963) Aust J Chem 16: 1056
158. Street GB, Clarke TC, Krounbi M, Kanazawa K, Lee V, Pfluger P, Scott JC, Weiser G (1982) Mol Cryst Liq Cryst 83: 1253
159. Street GB, Clarke TC, Geiss RH, Lee VY, Nazzal A, Pfluger P, Scott JC (1983) J Phys (Paris) Colloq 44: C3-599
160. Pfluger P, Krounbi M, Street GB, Weiser G (1983) J Chem Phys 78: 3212
161. Erlandsson R, Inganäs O, Lundström I, Salaneck WR (1985) Synth Met 10: 303
162. Erlandsson R, Lundström I (1983) J Phys (Paris) Colloq 44: C3-713
163. Neoh KG, Kang ET, Tan TC (1988) Polym Deg Stab 21: 93
164. Neoh KG, Kang Et, Tan KL (1991) J Appl Polym Sci 43: 573
165. Bätz P, Schmeisser D, Göpel W (1991) Phys Rev B 43: 9178
166. Neoh KG, Kang ET, Tan TC (1989) J Appl Polym Sci 37: 2169
167. Traore MK, Stevenson WTK, McCormick BJ, Dorey RC, Shao W, Meyers D (1991) Synth Met 40: 137
168. Wei Y, Hsueh KF (1989) J Polym Sci Part A, Polym Chem 27: 4351
169. Hagiwara T, Yamaura M, Iwata K (1988) Synth Met 25: 243
170. Neoh KG, Kang ET, Tan KL (1990) Polym Deg Stab 27: 107
171. Angelopoulos M, Ray A, MacDiarmid AG, Epstein AJ (1987) Synth Met 21: 21
172. Inoue M, Navarro ER, Inoue MB (1989) Synth Met 30: 199
173. Neoh KG, Kang ET, Tan KL (1991) Synth Met 40: 341
174. Andreatta A, Cao Y, Chiang JC, Heeger AJ (1988) Synth Met 26: 383
175. Li S, Dong H, Cao Y (1987) Synth Met 20: 141
176. Neoh KG, Kang ET, Tan KL (1991) J Phys Chem 95: 10151
177. Kang ET, Neoh KG, Tan KL (1992) Surf Interf Anal (In Press)
178. Kobayashi T, Yoneyama H, Tamura H (1984) J Electroanal Chem 177: 293
179. Stilwell DE, Park SM (1988) J Electrochem Soc 135: 2497
180. Österholm JE, Passiniemi P (1987) Synth Met 18: 213
181. Neoh KG, Kang ET, Tan KL (1991) Polym Deg Stab 31: 37
182. Tourillon G, Garnier F (1984) J Electroanal Chem 161: 407
183. Gustafsson G, Inganäs O, Nilsson JO, Liedberg B (1988) Synth Met 26: 297
184. Gustafsson G, Inganäs O, Nilsson JO (1989) Synth Met 28: C427
185. Nilsson JO, Gustafsson G, Inganäs O, Urdal K, Salaneck WR, Österholm JE, Laakso J (1989) Synth Met 28: C445
186. Veal BW, Paulikas AP (1985) Phys Rev B 31: 5399
187. Nilsson JO, Inganäs O (1989) Synth Met 31: 359
188. Fischer JP, Becker U, Halasz SP, Muck KF, Puschner H, Rosinger S, Schmidt A, Suhr HH (1979) J Polym Sci Polym Symp 63: 443
189. Ikada Y (1984) In: Dusek K (ed) Advances in Polymer Science 57, Springer, Berlin Heidelberg, p 104
190. Emi S, Murase Y, Hayashi T, Nakajima A (1990) J Appl Polym Sci 41: 2753
191. Kang ET, Neoh KG, Tan KL, Uyama Y, Morikawa N, Ikada Y (1992) Macromolecules 25: 1959
192. Briggs D, Brown A, Vickerman JC (1989) Handbook of static secondary ion mass spectroscopy, Wiley, New York
193. Wee ATS, Huan CHA, Gopalakrishnan R, Tan KL, Kang ET, Neoh KG, Shirakawa H (1991) Synth Met 45: 227

194. Everson MP, Helms JH (1991) Synth Met 40: 97
195. Mantovani JC, Warmack RJ, Annis BK, MacDiarmid AG, Scherr E (1990) J Appl Polym Sci
     40: 1693
196. Porter TL, Dillingham TR, Lee CY, Jones TA, Wheeler BL, Caple G (1991) Synth Met 40: 187
197. Drummond IW, Street FJ, Ogden LP, Surman DJ (1991) Scanning 13: 149

Editor: H. Fujita
Received February 1992

Author Index Volume 101–106

Subject Index